THE

GREAT

BURN

The Great Burn:

Wildfires

and

Their Growing Rate

Authors
Austin Mardon
Mujtaba Ansari, Arham Ansari, Aleefa Devji,
Brianna Bedran, Diana Eve Amiscaray, Amal Rizvi,
Daniel Klassen, JJ Doleweerd, Christina MacDonald,
Kelly Wu, Shanuga Thavarajah

Edited by
Tighe Andreou
Catherine Mardon

GM
★
PRESS

Cover Design and typeset by Clare Dalton
Free for use cover photograph by Sippakorn Yamkasikorn

Print ISBN 978-1-77369-896-0
Ebook ISBN 978-1-77369-897-7

Golden Meteorite Press
103 11919 82 St NW
Edmonton, AB T5B 2W3
www.goldenmeteoritepress.com

Chapters

Introduction

A spark, and fuel; that's all it takes to start a fire, and once a fire starts it can be hard to stop. The dangers from flames and smoke grow and blazes spread, often with an uncontrollable and unpredictable ferocity and unparalleled intensity. Through human history, fire has shaped our world, the harnessing of which is considered one of humanity's greatest accomplishments. But what happens when fire finds itself free of the shackles of human containment and is free to set the world ablaze? A wildfire is an uncontrolled fire that burns in wildland vegetation. While these fires often burn in rural areas, their impacts can be felt across communities, cities, and the globe.

In 11 short chapters, this book aims to identify and address the impacts that wildfires have across the globe. In Chapter 1, we identify the interlocking causes and impacts that wildfires have on North America. This chapter touches on the common reasons for wildfires within North America, before moving on to a discussion of the largest wildfires in North American history. This chapter finishes with a discussion of the impacts of these fires, the effects of climate change on said fires, and techniques for restoration and prevention. In Chapter 2, we move to South America; exploring how wildfires impact the continent as a whole. This chapter focuses on the impacts that wildfires have had on numerous notable countries in South America, including but not limited to, Brazil, Bolivia, and Argentina. Dive deep into the reasons behind these fires and the impacts that these fires have on the Amazon and its lifeforms. Chapter 3 expands our reach to Africa. Here, the reader will discover the connection between droughts and wildfires, as well as learn about the impact of burning upon the water cycle. Finally, the chapter concludes with an insight to the impacts of climate change and the effects that these fires have on African communities. In Chapter 4, we take a look at the causes and impacts of wildfires in Asia. This chapter focuses on the causes of wildfires within Asia and the impacts of smoke haze. Learn about wildfire hotspots and the causes and impacts of the Asian Forest Fires of 1997-1998. In Chapter 5, we continue our geographic survey with Australia, touching on how the island country's gorgeous landscape and diverse wildlife fall victim to uncontrollable blazes. This chapter touches on the impact of wildfires, both historical and present day, before expanding on fire management practices and the future of the country. Chapter 6 sees us round up our geographical survey with a trip to the Mediterranean. We take a look into the long-term effects of wildfires and how they are linked to humans. This chapter views the history of some of the greatest fires ever in the history of the Mediterranean region and how land-use changes and the shift to urbanization plays a role. In Chapter 7, we discuss the health effects of wildfire smoke. This chapter dives into a discussion of the health effect of wildfire smoke through a comparison between Alberta and California. We also discuss the mental health effects of wildfire smoke, and how the COVID-19 pandemic may affect public perceptions towards mitigation efforts including masking, and public awareness campaigns. In Chapter 8, we take a look at the monetary and health impacts of wildfires. This chapter identifies the impacts of wildfires on humans and their economic effects. Numerous notable wildfires and their effects are discussed and potential preventative measures are addressed. In Chapter 9, we address the possibility of investing in climate change and how global investment in combating climate change can lead to a potential reduction in wildfires. The impacts of climate change are viewed on a global scale, and a brief history of climate change activism is provided.

Costs, implications, and current actions are also discussed. In Chapter 10, carbon sources and carbon sinks are discussed. The role of carbon is illuminated through a robust analysis of carbon sources, sinks, and the carbon cycle. Readers will learn about a multitude of sources and sinks and how the carbon cycle plays a role in the world today. Finally, in Chapter 11, we address the impacts that wildfires can have on animal populations. This chapter touches on both the physical injuries from wildfires as well as the subsequent habitat loss that can occur for both terrestrial and aquatic life forms as a result of these uncontrolled blazes.

Chapter 1: Wildfires in North America

Mujtaba Ansari

Introduction

Updated as of December 2nd, 2022, according to statistics by the National Interagency Coordination Center (NICC), a total of 64.1 thousand wildfires have burned 7.3 million acres of land this year in the United States (CRS, 2022). The number of fires and the acres burned is predicted to keep increasing throughout North America, especially affecting Alaska, California, British Columbia, and Newfoundland and Labrador (Aziz, 2021). Wildfire prevention, emergency response, and recovery are highly expensive. Dealing with thousands of fires annually is an intensively challenging task for the wildfire management subsect. Due to its unpredictability, managing wildfires is extremely challenging. Even with advancements in predictive technology and tools, it cannot be exactly known when a wildfire will start, where it will start, how much damage it will cause, how long it will take to stop the fire, and how much the whole process would cost. The unpredictability and the growing rates of wildfires across North America continue to be a huge issue that will not slow down anytime soon (CRS, 2022).

Common reasons for wildfires in North America

Wildfires in North America are triggered and spread due to many situations and a collection of factors. The North American boreal forest, which stretches from Alaska to Newfoundland, is a common area for many small fires, and not as frequent, large fires. The infrequent large fires, which only make up 5% of the total fires that occur in the boreal forest, are responsible for over 85% of the area burned (Fauria and Johnson, 2007). The most common reason for wildfires in the boreal forest, and other areas with abundant plant life, is lightning. Lightning is a natural occurrence that cannot be prevented. However, lightning has caused wildfires since the beginning of time; the reason for the increase in the area burned throughout the years is caused by extended fire seasons due to climate warming (Fauria and Johnson, 2007). The second most common cause of wildfires, especially in areas near human settlements, is human activity. This includes people who go camping and leave their fires and potentially flammable equipment unattended, people who toss lit cigarettes and cigars, and people who commit intentional arson, are some of the ways that humans have increased wildfire rates (Moore, 2021). Finally, though rare, other natural occurrences can also trigger wildfires. These natural occurrences include volcanoes, meteors, and the sun (BIA, 2022). Forests and plant life in active volcanic regions easily catch on fire due to the heat, ash, and magma of a volcano. Meteors that hit forests and other grasslands can also spread wildfire very quickly due to the immense heat of the meteor. High temperatures and dry seasons can make the heat of the sun a cause of wildfires as well. Human involvement and natural occurrences have been the most common causes of wildfires in North America.

Largest wildfires in North American history

Wildfires can be very small. They can start, burn a small area, and can either extinguish themselves or be put out relatively quickly. However, some wildfires can be massive. They can start, spread very quickly, burn millions of acres, and take weeks to put out. Some of the largest wildfires, based on the amount of area burned, in North America, include the 2014 Northwest Territories fires, the Manitoba fires, the 2020 California fires, the 2011 Texas fires, and the 2018 British Columbia fires.

2014 Northwest Territories

The 2014 Northwest Territories summer season, which went from June to September, was the most severe wildfire season in the Northwest Territories history. The wildfires during the season burned over 8.4 million acres. A total of around 380 fires were reported throughout the boreal forest and surrounding lakes (Gaboriau et al., 2020). Lightning was reported to be the dominant firestarter; being responsible for around 95% of the wildfires. The large wildfires in the Northwest forced many indigenous tribes to leave their communities in order to stay safe from the fires and the health effects of the overwhelming smoke (Kochtubajda, 2019). Apart from the cost of damages, the firefighting cost alone was reported to be more than 56 million CAD (Gaboriau et al., 2020)

1989 Manitoba

Manitoba's 1989 wildfire season was recorded to be one of the most devastating in its history. The fires burned over 7.9 million acres of land in total. Out of all the acres burned in Canada during 1989, due to various wildfires, the Manitoba fires accounted for 43% of the total acres burned (Hirsch, 1991). The number of fires, which continues to be the highest record in Manitoba, came to around 1147 in total, with lightning and human activity being the main firestarters. The fires burned around 9% of the province's forest areas, forcing the evacuation of over 24 thousand people from the surrounding communities. The firefighting costs came out to be more than 68 million CAD (Hirsch, 1991).

2020 California

Known as the most "fire-prone" state in the United States, California has had many large wildfires in its history (Kerlin, 2022). Though not the biggest in California's history, the 2020 wildfires were the most expensive and the most recent. The 2020 fire season's total cost of overall losses came out to a staggering 19 billion USD. In addition, the firefighting cost reached 2.1 billion USD. The fire season caused the most wildfire economic losses in California's history (Kerlin, 2022). Around 10 thousand fires burned over 4.3 million acres in 2020, making it the first year that compared to the record wildfires in the 1800s, in regards to area burned in California. A total of 33 people were killed due to the fire, with more casualties as a result of its effects (Kerlin, 2022).

2011 Texas

A drought that started in October 2010 was at its peak by September 2011. It was also the worst drought in Texas history, making way for a destructive wildfire season. However, the state of Texas does not have a distinct season for wildfires, the risk of wildfires is present all year round, with some peak wildfire months (Jones et al., 2012). By the end of the 2011 Texas wildfire season, More than 4 million acres of land was burned by around 31 thousand wildfires. The causes being high winds, high temperatures, and dry conditions due to the drought. Around three thousand homes were destroyed but, with the help of firefighters around the United States, nearly 39 thousand homes were also saved. The fires also took the lives of four firefighters and six civilians (Jones et al., 2012). The total cost of firefighting and operation came out to be 216 million USD (Galbraith, 2011).

2018 British Columbia

Following a devastating record fire season in 2017, British Columbia faced an even greater fire season the following year. The 2018 fire season consisted of around two thousand fires. By the end of the 2018 season, the fires had burned over 3.2 million acres of land, while the

2017 fire season burned 2.9 million acres (Wang and Strong, 2019). The 2018 Wildfires in British Columbia made up 60% of the area burned in Canada that year. The total expense of firefighting and operations came to an estimated 615 million CAD. The 2018 and 2017 combined expenses resulted in British Columbia spending over 1 billion CAD on fire management. The estimated costs include the property damages (Wang and Strong, 2019). The Air Quality Health Index reported high health risks for weeks in the areas affected by the wildfires and smoke; resulting in children and elders with heart or breathing problems being especially vulnerable to the conditions (Wang and Strong, 2019).

Impacts on human health

Global warming continues to be a threat to the environment and human health. These temperature changes lead to devastating natural disasters such as wildfires. The changes in climate are linked to the steady increase in wildfire intensity, frequency, and duration (Rossiello and Szema, 2019). Wildfires cause exceptional economic losses as well. Year by year, wildfires reach new records of area burned and economic losses in North America, despite prevention efforts. The costs include property damages, structure damages, operations costs, firefighting costs, aircraft costs, and loss of resources such as water, lumber, and vegetation (D'Evelyn et al., 2022).

However, economic losses and the destruction of structures are not the only negative impact of wildfires. Many studies have found that exposure to smoke from wildfires increases the risk of respiratory and cardiovascular illnesses. Various studies have also found that exposure to wildfire smoke can negatively impact childbirth. Individuals with existing respiratory illnesses, vascular illnesses, and diabetes, are especially at risk of health impacts from wildfire smoke. When exposed to smoke, children are at a higher risk of developing asthma, lung development issues, and improper lung functions (D'Evelyn et al.,

2022). Wildfire smoke is dangerous because it contains a list of air pollutants such as nitrogen dioxide, carbon monoxide, and volatile organic compounds, all of which are a major concern to public health (Rossiello and Szema, 2019).

Populations with higher exposure to the toxic smoke include low-income families with poor air filtration in their houses, indigenous communities, homeless people, and individuals with high-exposure work. The worst health impact from wildfires results from direct contact with the fires. These injuries could be minor to severe burns. However, though not very common, firefighters and civilians can and have lost their lives due to being burned by the wildfires or choking on the overwhelming smoke (D'Evelyn et al., 2022).

Wildfire damage in North America in recent years

Statistics have found that the number of wildfires in North America has slightly decreased when compared to the past three decades (CRS, 2022). However, the number of acres burned due to wildfires has slightly increased. In 2017, a total of 71.5 thousand fires burned ten million acres. In 2018, a total of 58.1 thousand fires burned around 8.8 million acres. In 2019, a total of 50.5 thousand fires burned 4.7 million acres of land. In 2020, a total of 59 thousand fires burned around ten million acres of land. In 2021, a total of 59 thousand fires burned over seven million acres of land. Finally, as of December 27, 2022, around 64 thousand fires have burned over 7.3 million acres this year in North America (CRS, 2022). The average number of annual wildfires from 2017 to 2022, comes to around 60 thousand, while the average number of acres annually burned comes to around 7.9 million. In comparison, from the year 2000 up until now, the average number of annual wildfires is around 70 thousand, and the average number of acres annually burned is around seven million. This comparison between these past years shows that the amount of acres burned is increasing (CRS, 2022).

The number of wildfires that take place within a year and the area that they burn do not exactly represent the damage and impact on human settlement and wildlife/plant life. Many wildfires occur in large, unpopulated, and undeveloped areas. Though the fires might affect businesses such as logging, the economic losses are relatively low. This goes to show that the number of acres burned does not equal economical damage. The number of fires and acres burned also does not relate to the intensity of the wildfire. The intensity of the fires is related to the ecological impacts. The wildfire could destroy the soil. The soil would not be able to grow any plants, making it useless for agriculture and negatively affecting the wildlife that lived or depended on that area. Ecological impacts could also be seen on a wider, forest scale. A severe wildfire can affect the whole forest rather than just the area that was burned by affecting the root systems, underground water, and various nutrients required for the forest's health (CRS, 2022).

Predicting the damages of a fire can be quite difficult. Out of the over 1.5 million fires that have occurred since 2000, only 237 fires were responsible for burning over 100 thousand acres, while only 15 were able to burn more than 500 thousand acres. A fire's potential to be catastrophic depends on factors such as geography and weather. Between 2017 and past 2021, around 1065 fires have been classified as catastrophic, which surpasses the general statistical 1% of fires that become significant. Hence, the statistical data shows that though the number of annual wildfires in recent years in North America has decreased, the chances of those fires becoming significant and burning excessive acres have increased (CRS, 2022).

Wildfires and climate change

The constant and significant changes that have impacted the North American landscapes are primarily due to three factors. These factors include the build-up of fuels, the growing urban-wildland development, and the drying and warming climate. The changes in climate have undoubtedly been a significant factor (Schoennagel et al., 2017). The changes in wildfire activity have a close relation with climate change, which can be seen throughout the years. Observed from 1979 to 2013 on a global scale, the normal fire season length has increased by an average of 19%, with the western United States having some of the most notable increases. Though the increase in the amount of area burned in North America is not solely because of climate change, it is responsible for over 50% of the increase. The increase in area burned and occurrences of wildfires in the forest regions of North America, are a result of longer fire seasons, early snowmelt, rising temperatures, and increased droughts (Schoennagel et al., 2017).

A way that climate change can make wildfires worse is by increasing the number of ignitions. The probability of ignitions is bound to increase in warmer temperatures because as the temperature gets warmer, cloud-to-ground lightning also increases (Flannigan et al., 2000). Along with increased ignitions, climate change is predicted to increase the fire season length further. It has been predicted that fire seasons will start during early spring and continue longer, well into the autumn season. In Canada alone, the fire seasons are predicted to increase by an average of 22% or 30 days in correlation with increased CO_2 climate. In order to better understand the impact of climate change on the frequency and length of wildfires, scientists simulated a climate that has two times the amount of CO_2 than now, while the rest of the factors that make up wildfires stay the same. The researchers found that in the United States, the frequency of lightning fires increased by 44%, and the amount of area burned increased by 78%, in a double CO_2 amount scenario.

Simulations performed in Canada also showed a significant increase in fire severity. Additionally, the results showed an increased overall dryness, though not as significant (Flannigan et al., 2000). In the double CO_2 scenario, some places in North America and Europe actually showed less fire severity due to the increase in precipitation, showing that climate change and warmer temperatures are not solely responsible for increases in wildfires (Flannigan et al., 2000).

The correlation between climate change and increased wildfire activity goes both ways. Wildfire activity is also a driver of climate change (Liu et al., 2016). The smoke resulting from wildfires is packed with greenhouse gasses, pollutants, and various particles. Under normal circumstances, these pollutants and greenhouse gasses released into the environment are removed by new plant life that grows in the burned areas. However, the increase in fire severity, fire season, and the number of fires, does not allow for all the pollutants to be removed. These pollutants, especially carbon dioxide and methane, increase the overall CO_2 and lower the air quality in a given environment. These environmental changes can lead to short-term and sometimes long-term climate change effects (Liu et al., 2016). The increase in wildfires due to climate change leads to wildfires being one of the factors leading to climate change.

Post-wildfire restoration and prevention

After the fires are put out, the fire season is almost at an end, and damages are being repaired, the post-fire restoration and prevention must begin. Affected forests are dangerous for people and uninhabitable by wildlife. There are possibilities of mudslides, water contamination, and loss of wildlife habitat (Hallaman, 2020). The first step is to prioritize areas that need immediate attention, such as areas that require erosion control. The next priorities are given to areas in terms of high, medium, and low burn severity, since high-severity burned areas will require more help. The second step would be to seek out local experts which can advise and give detailed information. The experts could provide information regarding the wildlife species, tree species, and geographical data. Using the info gained from experts, the restoration service can begin to grow the necessary seedlings for the restoration. The seedlings will need to be determined fast because their required growth takes up to a year and sometimes more. The third step would be to gather volunteers, funds, agencies, and other partners in order to smoothly execute the restoration plans. The final step is to follow the planting plan for several years, all while monitoring and managing long-term forest health (Hallaman, 2020). Prevention measures for future wildfires must also be taken. After a fire, the affected areas are left with dry bushes and grass - the optimal fuel for the next wildfire. The removal of these bushes and grass take priority in the prevention plan. Another step of the prevention plan could be to thin out the densest and most fire-prone parts of the forest (Hallaman, 2020).

Conclusion

North America's boreal forests and western United States areas are highly prone to wildfires and have a history of devastating fire seasons. The ignitors of these fires could be natural occurrences such as lightning or caused due to human activities. Some wildfires have reached an astounding amount of area burned. For example, the 2014 Northwest Territories fire season. Although there are many economical losses and structural destruction caused by wildfires, human health is also put at risk. The smoke from the fires could worsen or cause respiratory and cardiovascular illnesses, especially in kids and the elderly. In comparison with wildfires in past years, recent years have had more total acres burned with a lesser number of recorded fires. A reason for the increase in fire severity over the years could be climate change. Even though wildfires have been predicted to get worse in future years with limited options to stop the increase, developing restoration and prevention efforts could be a strong response to this growing problem.

References

Aziz, S. (2021, July 21). *A look at Canada's wildfires in numbers and graphics over the decades.* Global News; Global News. https://globalnews.ca/news/8045796/canada-wildfires-yearly-trends/

Congressional Research Service. (2021). *Wildfire Statistics.* https://sgp.fas.org/crs/misc/IF10244.pdf

D'Evelyn, S. M., et al. (2022). Wildfire, Smoke Exposure, Human Health, and Environmental Justice Need to be Integrated into Forest Restoration and Management. *Current Environmental Health Reports, 9*(3), 366–385. https://doi.org/10.1007/s40572-022-00355-7

Flannigan, M. D., Stocks, B. J., & Wotton, B. M. (2000). Climate change and forest fires. *Science of the Total Environment, 262*(3), 221–229. https://doi.org/10.1016/s0048-9697(00)00524-6

Gaboriau, D. M., Remy, C. C., Girardin, M. P., Asselin, H., Hély, C., Bergeron, Y., & Ali, A. A. (2020). Temperature and fuel availability control fire size/severity in the boreal forest of central Northwest Territories, Canada. *Quaternary Science Reviews, 250*, 106697. https://doi.org/10.1016/j.quascirev.2020.106697

Galbraith, K. (2011, September 8). *Texas Politicians Press Feds for Fire Relief Money.* The Texas Tribune; The Texas Tribune. https://www.texastribune.org/2011/09/08/texas-firefighting-funding-woes-mount/

Hallaman, J. (2020, October 21). *Wildfire Restoration 101: How to Drive Recovery and Prevent Future Disaster*. Arbor Day Blog. https://arbordayblog.org/wildfire-recovery/wildfire-restoration-101/#:~:text=Wildfire%20restoration%20is%20a%20long,by%20the%20disaster%20step%20in.

Hirsch, K. G. (1991). A chronological overview of the 1989 fire season in Manitoba. *The Forestry Chronicle, 67*(4), 358–365. https://doi.org/10.5558/tfc67358-4

Kerlin, K. E. (2022, May 4). *California's 2020 Wildfire Season*. UC Davis. https://www.ucdavis.edu/climate/news/californias-2020-wildfire-season-numbers

Kochtubajda, B. (2018, December 12). *An Assessment of Surface and Atmospheric Conditions Associated with the Extreme 2014 Wildfire Season in Canada's Northwest Territories*. Taylor and Francis Online. https://www.tandfonline.com/doi/abs/10.1080/07055900.2019.1576023?journalCode=tato20

Liu, J. C., et al. (2016). Particulate air pollution from wildfires in the Western US under climate change. *Climatic Change, 138*(3-4), 655–666. https://doi.org/10.1007/s10584-016-1762-6

Macias Fauria, M., & Johnson, E. A. (2007). Climate and wildfires in the North American boreal forest. *Philosophical Transactions of the Royal Society B: Biological Sciences, 363*(1501), 2315–2327. https://doi.org/10.1098/rstb.2007.2202

Moore, A. (2021). *Explainer: How Wildfires Start and Spread*. College of Natural Resources News. https://cnr.ncsu.edu/news/2021/12/explainer-how-wildfires-start-and-spread/

Rossiello, M. R., & Szema, A. (2019). Health Effects of Climate Change-induced Wildfires and Heatwaves. *Cureus*. https://doi.org/10.7759/cureus.4771

Schoennagel, T., et al. (2017). Adapt to more wildfire in western North American forests as climate changes. *Proceedings of the National Academy of Sciences*, *114*(18), 4582–4590. https://doi.org/10.1073/pnas.1617464114 *2011 Texas Wildfires Common Denominators of Home Destruction*. (2012, September 4). TexasA&M. https://tfsweb.tamu.edu/uploadedFiles/TFSMain/Preparing_for_Wildfires/Prepare_Your_Home_for_Wildfires/Contact_Us/2011%20Texas%20Wildfires.pdf

Wang, J., & Strong, K. (2019, May 29). *British Columbia's forest fires, 2018*. Statistics Canada. https://epe.lac-bac.gc.ca/100/201/301/weekly_acquisitions_list-ef/2019/19-22/publications.gc.ca/collections/collection_2019/statcan/16-508-x/16-508-x2019002-eng.pdf

Wildfire Investigations | Indian Affairs. (2022). Bia.gov. https://www.bia.gov/service/wildfire-prevention/wildfire-investigations

Chapter 2: Wildfires in South America

Arham Ansari

Wildfires in South America Recorded Activity in 2022

According to National Geographic : "A wildfire is an uncontrolled fire that burns in the wildland vegetation, often in rural areas (World Health Organization. (n.d.). Wildfires can burn in forests, grasslands, savannas, and other ecosystems, and have been doing so for hundreds of millions of years. They are not limited to a particular continent or environment. Wildfires can burn in vegetation located both in and above the soil. Ground fires typically ignite in soil thick with organic matter that can feed the flames, like plant roots. Ground fires can smolder for a long time—even an entire season—until conditions are right for them to grow to a surface or crown fire. Surface fires, on the other hand, burn in dead or dry vegetation that is lying or growing just above the ground. Parched grass or fallen leaves often fuel surface fires. Crown fires burn in the leaves and canopies of trees and shrubs." South America has been experiencing unusually high wildfires since the beginning of 2022. Last year the highest numbers of wildfires were reported with a total of 37,934 fires in South American countries (World Health Organization. (n.d.). A hot spell and drought conditions across north Argentina and Paraguay have recorded wildfires in the region that cause higher fire emissions and smoke, Paraguay put estimated carbon emissions from fires around 5 megatons indicating exceptional wildfire activity. The record shows that Northeastern provinces like Santa Fe, Formosa, and Misiones also experience high emissions. the condition in Paraguay started to improve at the end of February. Other areas like Colombia and Venezuela have been exposed to fire as the spring season started and faced wildfire emissions related to drought and increasingly dangerous wildfire conditions in the Amazon and also Orinoco Valley. South American countries have been facing a drought situation that increased wildfire activity in the whole region (World Health Organization. (n.d.). In 2022, wildfire numbers are at their peak since 2010, when the Amazon faced severe dry weather and warm ocean water of the North Atlantic. The world's largest wetland suffered from more fires in 2022 than any previous records. An ecologist of The Amazon Environmental Research Institute (IPAM), said if there is another drought this year, the condition will be worse than before (World Health Organization. (n.d.). Satellites recorded 32,017 fire spots in the world's largest rainforest in September which is a 61% increase in growth from September 2019. These fires not only burn the deforested lands but also burn the virgin forest which shows the rainforest is becoming parched and more inclined to fire (World Health Organization. (n.d.). An analysis by a federal university found that 23% of the wetlands, which are the accommodation of the large population of Jaguars in the whole world, have burned (World Health Organization. (n.d.).

Wildfire in South America

The Copernicus Atmosphere Monitoring Service (CAMS) has shown recorded wildfire intensity through satellite observations of fire radiative power (FRP) in South America, especially Brazil, Paraguay, Argentina, Colombia, and Venezuela and they faced the longest and most destructive environmental crisis by these neighboring countries. CAMS Global Fire Assimilation System (GFAS) inquiry shows the fact that fire activities in Paraguay and Argentina between January and February touched the record of carbon emission since 2003 (South America sees record wildfire activity in early 2022, n.d).

Brazil

The peak season of wildfire in Brazil normally begins in august and continues for around 14 weeks. In 2021, throughout the year Brazil outlined around 184 thousand wildfire outbursts which is the highest number in South America. In the last 4 weeks, Brazil lost the most significant burned area called Rio Grande do Sul, which means 3.8% of the land burned in Brazil and that is unfortunately very high as compared to the period going back to 2001. The second greatest figure reported in Bolivia at over 34000, In French Guiana 88 wildfire outbreaks were reported (Vizzuality. (n.d.).

Bolivia

According to NASA, in early September a wildfire started and burned 5% of Tucabaca valley. Tucabaca was already severely burnt in 2019 and 2021, during those two years around 33.3 square kilometers of forest were lost and damaged the wildlife habitat. In September, Bolivia was declared a state of emergency, and the temperature touched 45C (Vizzuality. (n.d.).

Otuquis National Park, located in the south of Tucabaca along with the border of Brazil and Paraguay was also afflicted by wildfire and spread around 80 kilometers for nearly two weeks. Furthermore, Otuquis is also a part of an important wetland area (RAMSAR), it is populated by many species like marsh deer, giant otters, wolves, and Bolivian anacondas that are heavily affected by wildfires. Along with the loss of habitat, fire smoke brought serious health issues like vision and respiratory problems to the closer residents. Despite the fact that controlling the wildfire in Otuquis National Has been fortunately achieved, the fear of fire reignition remains (Vizzuality. (n.d.).

Argentina

In Argentina, most of its provinces are facing the most severe wildfire in decades, causing eye infections and respiratory issues while burning wild animals- from monkeys to jaguars, as well as wild birds and reptiles (Vizzuality. (n.d.).

NASA satellites reported a wildfire in Argentina set on fire on December 22 and lasting for several weeks, has burned approximately 1 million acres of land which is higher as compared to the previous year in 2001. During 4 weeks of statistics in Argentina, the most notable affected area was Corrientes, which represents 76% of the total land burned in Argentina and it is unexpectedly high as compared to the same time back in 2001. The peak season of wildfire in Argentina starts in August and lasts for around five months. From 1002-2021, they lost 535 kha of land because of wildfire & due to other reasons, they lost 5.7Mha. ROSARIO, Argentina Grassland fires near the delta river in south America cause dangers to wetlands ecological community and human health.

South America's second-largest water source "The Parana River" drop-down to its lowest level in 77 years due to the ongoing drought. High temperatures, gales, and low humidity create an ideal situation to begin wildfires.

Paraguay

Paraguay has been affected because of changing weather trends in the EI Nino, which is the big reason for severe drought and flood. Extreme hot weather and deforestation caused wildfires around the country that increased the risks in the central north province of Concepcion, Amambay, and some areas of San Pedro and Canindeyu. An emergency was declared at the national level as 12,000 outbreaks were detected and the capital, Asuncion heavily affected by dense smoke. Wildfires have been extremely extensive in the Gran Chaco Forest. The local mayor of

Chaco town says " uncontrollable wildfires" affected hundreds of miles. In 2019, destructive wildfires damaged 320,000 hectares of the covered area with forest in Paraguay. A further 150,000 hectares of forest were destroyed in 2020. Over 25000 wildfire blazes were marked in 2021. In less than a month, wildfires spread over 240,000 acres of land in Paraguay which border Brazil and Bolivia. The world's largest tropical wetland area, and the world's largest flooded grassland "the home of extensive biodiversity" is also under great threat. A report showed the fact that Paraguay is one of the countries where forests are permanently lost after wildfires (Vizzuality. (n.d.).

Columbia
During 2001 to 2021, Colombia vanished its natural forest of around 331Kha from wildfires to extend 72% of its land for agriculture and 4.60Mha from other sources of loss. The most forest loss due to wildfires was in 2016 with 49.3Kha (17%). In Colombia, the peak time period of wildfires begins in January and lasts for around 13-14 weeks. Around 2,099 fire alerts were reported in just one year's time period (Dec 2021- Dec 2022). In Colombia, the most recent four weeks of data show that the most crucial destroyed area was Norte de Santander where 433 ha of land was burned (Vizzuality. (n.d.).
In 2021, approximately 13.6 thousand wildfire outbreaks were detected throughout Colombia. The region in Colombia named Antioquia had the highest rate of forest loss due to wildfires during 2001-2021 with an average of 1.98kha burned per year.

Reasons Behind the Wildfires

Approximately more than 85% of wildfires in South American countries are caused by humans (Oliveira et al., 2022). The heart issue of wildfires in Brazil is Amazon's deforestation for farming land and industrial cultivation is one of the notable drivers of deforestation from the first six months of the current year, 6800 square kilometers were cleared for farming. The Amazon's deforestation has been continuously increasing for most of the time in a decade which is due to the country's agricultural objective to become the major producer in the global market. Last year, the deforestation rate of Amazon in Brazil elevated to the highest extent in 15 years (Oliveira et al., 2022). In Brazil, livestock farmers and landowners place the Amazon on fire to clear land and expand business illegally. The land is being destroyed through wildfire to create a system for livestock raising. The Savannah is a giant farming field for soybean production even though it is legally not allowed in the Brazilian Amazon and Brazil is the largest producer of soybean in the world. Millions of hectares of brail and Argentina have been used for the plantation of soybeans There are several natural reasons for wildfires are severe weather conditions, and droughts (Oliveira et al., 2022). The destruction of the Amazon, the Cerrado, and the Pantanal has affected indigenous people's land, and they striving to protect their forests and territories for years. These invasions are illegal and a violation of human rights. Indigenous lands are the best option to save forests (Oliveira et al., 2022). Repeated wildfire activities in South America is one the most concerned event in recent years which impact the local ecosystem as well as the global climate. The blazed area and fire emissions are predicted to rise in coming future because of high temperatures and drought conditions. In South America, Brazil experienced the highest level of wildfires mostly affecting Brazil's Cerrado region, where the climate situation is especially hot and dry, especially from June- October season. However, expanded wildfires have historically affected large areas of land, not only in the Cerrado region but also in the Amazon and pantanal. For the very first time it is recorded that the wildfire smoke from Amazon reached São Paulo where around 2.7000 kilometers was burned in 2019, and in 2020 1/3 of the Pantanal biosphere was destroyed. Moreover, wildfires in the Cerrado and Pantanal regions have a higher spread rate than fires in Amazon and Atlantic forests (Oliveira et al., 2022). It is previously documented that wildfires have not been a major problem across the Amazon Forest as it has a higher wet and natural climate. Though, the dangers of wildfires may rise over a broad spectrum in the next century due to severe climate changes and drought conditions. Presently, around 58% of the Amazon rainforest is severely sultry that contribute to wildfires and it is estimated that climate changes may reduce to 37% by 2050 (Oliveira et al., 2022). Wildfires in this region are primarily triggered by humans for farming land and deforestation activities, land use is established and tend to increase for soy production, on the other hand, the wildfires events are the result of poor environmental administration as well as Government has not introduced any additional effective public policies to fight against wildfires.

The weak environmental policies rising the deforestation that tend to intensify wildfires events in upcoming years.Relationship between climate, land farming, and wildfires are complex and dependent and deadly affect wildlife habitat, destroying houses and heavily polluted air (Oliveira et al., 2022). Most of the research work has been completed in Brazil around 74 studies, Argentina 64 followed by other regions of South America. Scientists who worked closely about the factors like fire intensity and rate of spread have come to know the extensive blend of knowledge. Ecologists acknowledge that most of the wildfires are naturally occurring but contribute to the community's vulnerability across all areas to support forests and surrounding communities (Oliveira et al., 2022).

Dense smoke generated from wildfires is the major concern of public health sectors (US Forest Service. (n.d.)). Infants, children, pregnant women and elderly people are most vulnerable to short term as well as long term health concerns that generate from wildfire smoke and ashes. Moreover, firefighters and emergency responders are also greatly affected as they work closely in such areas. Major and minor injuries, burns and breathing smoke that cause eye, nose, throat infections.

Other diseases like bronchitis, Asthma, cardiac arrest and respiratory issues are caused by wildfire smoke in nearby communities. Wildfires also discharge a great notable quantity of mercury into air which may cause diseases such as loss of speech, hearing impairment, walking disorders and vision issues for all age groups (US Forest Service. (n.d.). A framework named " blue sky smoke modeling " setup the broad nationwide smoke forecast systems and developed smoke map and also shares current air quality and wildfire destinations across South America. Mapping of blazed areas helps to figure out the historical details of wildfires, so the most destructive lands have a history of intensity, repetitive fires, expanded blazed land per fire activity, less stretch between active wildfires and absolute fires in dry season. Mapping may contribute to developing meaningful policies to control this environmental crisis (US Forest Service. (n.d.). Wildfires in South America greatly affected native grassland that directly led to the physiological pressure and anxiety of local people. To roughly calculate these impacts, NDVI (Normalized Difference Vegetation Index)were derived that specify the grassland's photosynthetic activity and leaf density as an agent of its physiological health on a monthly basis. It is absolutely worthy to mention two successful programs " The Fire Mitigation Program of Alianca da Terra "that accomplished prevention and fire brigade campaigns in public as well private lands. Another program named " The innovative fire spread risk and monitoring system was established by the Federal University of Minas Gerri's and INPE 9 National Institute for Space Research (US Forest Service. (n.d.). The results from different research indicates that wildfires are likely to be more common and severe over the few centuries with equal impacts. Thus, availability of firefighters and resources will become more complex and expensive. While urgent global action plans and policies will be needed, to reduce greenhouse emissions remain important to stop escalating wildfires (US Forest Service. (n.d.).

How do the Amazon Wildfires Affect Wildlife?

The Amazon rainforest is the homeland of 30% of the world's species including jaguars, anteaters and birds that live in 160,000 sq km area and most biodiverse land in the whole world (Gill, 2021). Around 9,000 wildfires continuously raging all around the forests of Brazil, Bolivia, Paraguay, Colombia, and Peru. Scientists in South America estimated that around 17 million vertebrates such as reptiles, wild birds, snakes and rodents died in the Pantanal region that covers Brazil, Bolivia, and Paraguay. Researchers managed to reach affected areas of wetlands and inspect every dead animal and they identified 300 species of animals that were found dead. Moreover, an extremely wide number of snake species were killed due to wildfires (Gill, 2021). In fact Amazon's biodiversity provides a variety of beneficial impacts for humans, like special plants that are widely used in traditional medicines and support pharmaceuticals that treat patients of diabetes and Alzheimer's diseases. Trees and plants also benefit the atmosphere by absorbing carbon emissions. Wildlife in the South American region is the great means of support and food security for the local community. Farmers of the Xingu region in northern Brazil, working harder to restore their forest through traditional style called "Muvuca"- in which the native people sow the variety of seeds like cashew and açai and already planted 1.8 million trees (Global Fire Monitoring Center. GFMC. (n.d.))

Researchers utilize data of around 15,000 plants and vertebrate families to create biodiversity maps of the Amazon region. They found that around 40-73 thousand square miles of Amazon have been affected by wildfires since 2001. Almost 95% of all species living in Amazon rainforest were affected as well as around 85% of species are threatened (Global Fire Monitoring Center. GFMC. (n.d.))

The South American government in most areas does not provide sufficient financial support to develop and execute national fire management plans except for Argentina, Brazil and Chile (Global Fire Monitoring Center. GFMC. (n.d.)). There are no clear assigned responsibilities to the fire management department. The professional skilled workers for wildfire prevention are minimal and don't have wildfire suppression potential which doesn't allow the implementation of fire management plans and procedures (Global Fire Monitoring Center. GFMC. (n.d.))

Guidelines

Directives: check with community officials for local directives concerning wildfires risks and dangers. To ensure that project planning regulations, possible building rules, existing plans for cautions and evacuation, any national rules & regulations (Paraguay, n.d.).

Protection: wildfire damages and deprivations protection should be an essential part of policy. Work with a team and discuss multiple risk mitigation intended to process, which might result in a more customized insurance policy (Paraguay, n.d.).

Avoid Hazards: Man Made fire activities and functioning of certain machinery can raise the potential for wildfire combustion. In most regions, these activities include welding, grinding, faulty machinery and campfires. Projects that involve retention and utility of hazardous materials should be extensively managed. Understanding of the utility of local wildfire weather warning systems and their trigger levels to avoid fire ignition activities and keep the community safer (Paraguay, n.d.).

Risk Reduction: Wildfire risk reduction can be successfully achievable through a combination of strategies which includes location, site design, sources management and quick emergency response. Fuel reduction is one of the key factors for managing the wildfire hazards (Paraguay, n.d.).

Weather awareness and preparedness: wildfire weather potentially determined by a blend of strong wind, high temperatures, humidity and intensity of rainfall history. Weather index indicates dryer and more combustible fuels that will drive wildfires. It can widely spread through areas of grassy or wood fuels. Regional firefighting department is a potentially useful resource to know the detailed wildfire history of the regions (Paraguay, n.d.).

References

Gill, V. (2021, December 16). Brazil wildfires killed an estimated 17 million animals. BBC News.

Giorgis, M. A., Zeballos, S. R., Carbone, L., Zimmermann, H., von Wehrden, H., Aguilar, R., Ferreras, A. E., Tecco, P. A., Kowaljow, E., Barri, F., Gurvich, D. E., Villagra, P., & Jaureguiberry, P. (2021, August 9). A review of fire effects across South American ecosystems: The role of climate and time since fire - fire ecology. SpringerOpen.

Global Fire Monitoring Center. GFMC. (n.d.). Oliveira, U., Soares-Filho, B., Bustamante, M., Gomes, L., Ometto, J. P., & Rajão, R. (2022, January 28). Determinants of fire impact in the Brazilian biomes. Frontiers.

Pacific Northwest Research Station. Fire Effects on the Environment | Pacific Northwest Research Station | PNW - US Forest Service. (n.d.).

Paraguay. Think hazard - paraguay - wildfire. (n.d.).

Price, K. (2021, September 1). Study: How years of wildfires have devastated the Amazon. Conservation International.

South American fires and their impacts on ... - Wiley Online Library. (n.d.).

South America sees record wildfire activity in early 2022. Homepage. (n.d.).

Vizzuality. (n.d.). Brazil deforestation rates & statistics: GFW. Global Forest Watch.

World Health Organization. (n.d.). Wildfires. World Health Organization.

Chapter 3: Wildfires in Africa

Aleefa Devji

Wildfires in the African subcontinent make up a large amount of the world's burned land mass. In Africa, 22% of the continent burns every 8 years and it is estimated that approximately 9% of the land will burn on an annual basis (De Sales, et al., 2019). There are many causes of wildfires, such as air temperature and precipitation that are in line with climate change theories that affect wildfire significance as well as the average fire danger and frequency (Wanzala & Ogallo, 2020).

Although many fires occur by natural phenomenon, such as lightning strikes or volcanic eruptions, the majority of fires that have occurred in Eastern Africa over the past few years have been as a result of human activity that is mostly linked to agricultural processes. Some such activities linked to agricultural processes include the *slash and burn* practices that are used to prepare the land for crop planting or after crop harvesting has been completed (Styger, et al. 2007). Other activities include pastoral activities such as clearing land for livestock or for regeneration of pastures, as well as honey collection and charcoal burning. These man made fires that are initiated by people who use the land for agricultural means sometimes lead to uncontrollable spread, and can lead to large amounts of land destruction, destruction of properties, the losses of human lives, as well as natural habitats and biodiversity loss.

The Connection Between Drought and Wildfires

A study by NASA scientist Reiny in 2017 has found that there is a significant connection between drought and wildfires in sub-Saharan Africa. Although many wildfires are the cause of climate change, the research has been able to show how human farming activities associated with burning of land has affected the water cycle and inadvertently created a larger risk for wildfires.

Drought is not an new phenomenon experienced by northern sub-Saharan African, but in recent years the water shortages have been most detrimental in the Sahel - a region of land south of the Sahara Desert and stretching across the entire continent. Most recently in 2012, the drought struck and triggered food shortages so widespread that millions of people were suffering as a result of the crop failure and subsequent soaring food prices (Reiny, 2017).

Although there are various factors that influence the droughts in Africa such as periodic temperature shifts over the Atlantic Ocean, the overgrazing of the land also plays a crucial role. Overgrazing reduces vegetative cover and the ability for soil to retain moisture which would normally help contribute water vapor to the atmosphere in order to help generate rainfall. Instead, the bare and shiny desert soil that lacks the moisture is only able to reflect sunlight back into space, and has no capacity to increase rainfall (Reiny, 2017).

Overgrazing of the land is not the only human-caused culprit that has led to drought over the past decade, but biomass burning is another. Herders burn land to stimulate the growth of grass, and farmers burn landscapes to convert terrain to farming land as well as to get rid of unwanted biomass after the harvest season. These activities lend to similar consequences as overgrazing of the land because the fires themselves dry out the soil and dampen the ability for convention to bring rainfall

(Reiny, 2017). As well, aerosolized particles that enter the air by smoke caused by fire may also reduce the likelihood of rainfall. This happens because it causes a change in the sizes and types of aerosols in the atmosphere, adding pollution to the air, and disturbing the natural ability for cloud cover to form. Normally, water vapor in the atmosphere condenses on certain types and sizes of aerosol particles and creates cloud condensation nuclei in order to form clouds. Now as a result of the aerosols entering the air from smoke, we have too many aerosols in the atmosphere and water vapor is spread out more diffusely to a point where dense cloud cover is unable to form and therefore raindrops are not able to materialize.

NASA's Look at The Impacts of Burning on the Water Cycle

These effects of overgrazing and the associated activities of burning land mass or biomass on drought and the lack of overall rainfall has led researchers such as Charles Ichoku, a senior scientist at NASA's Goddard Space Flight Centre in Greenbelt, Maryland to further the research on the impacts of burning on the whole spectrum of the water cycle. Ichoku and his colleagues were able to use satellite records from 2001 to 2014 that included data from Nasa's Tropic Rainfall Measurement Mission to analyze the impact of fires of various water cycle indicators. Some of these indicators include soil moisture, precipitation, evapotranspiration, and vegetation greenness. Their research also went further to look into the interactions between clouds and smoke and the effects of fires on surface brightness.

The satellite data was set to match fire activity to hydrological indicators, and a pattern emerged and outlined that there is a direct relationship between fire and precipitation. When Ichoku looked at this data and the trends, he said, "there is a tendency for the net influence of fire to suppress precipitation in northern sub-Saharan Africa (Reiny, 2017)."

What this means is that human activity involved in farming practices and overgrazing as well as burning of crop fields post harvest, or burning of land and landscapes has directly contributed to the droughts and has thereby increased the propensity and likelihood for wildfires resulting from the dry climates and terrain. For example, the data revealed that in years that had more than average burning during the dry season, measurements of the soil moisture and evaporation decreased in the following winter season.

Climate Change on the Intensity of Wildfires in Africa

In the African subcontinent region there are two fire seasons, September to March in areas north of the equator, and April to August in areas south of the equator (Kayijamahe & Otieno, 2020). As previously mentioned, these fire seasons coincide with the farming and harvesting times where farmers engage in practices to cut down some of their vegetation and set fire to the rest in order to clear land for planting of crops. These practices are used by farmers as they are considered the cheapest and have the best advantage for killing off pests and diseases while also providing the best nutrients to the soil for future crops. However, this technique of burning is quite controversial due to its impacts on deforestation, soil erosion, and loss of biodiversity (Kayijamahe & Otieno, 2020). An example of the loss of biodiversity that has stemmed from this technique is linked to the Mount Kenya and Aberdare forests where key water towers and biodiversity hotspots were set on fire during an attempt to clear lands for farming, but let to the destruction of thousands of hectares of land and the killing of mass amounts of biodiversity.

Another area that is largely responsible for fires in Africa is South Sudan, where the fire season is November to March (Kayijamahe & Otieno, 2020). A study completed by GMES and Africa project showed that following a fire season if there is a good rainfall season, abundant vegetation is able to develop although this once again provides further abundance of biomass which leads to the consequential increase in fire incidences. The reason that this is of importance to note is that these fires can lead to wildfires, also known as wildland fire, bushfire or forest fire that can be uncontrolled in an area. Wildfires can arise from uncontrolled use of fires by untrained professionals who utilize burning for crop yield and grazing which result in detrimental losses of land, destruction of properties, as well as the loss of human lives (De Sales, et al., 2019). These fires have significant side effects that are different from controlled fires that are used by professionals and are set intentionally to pre-emptively prevent more serious wildfire by burning fuel for the fire early on before it can accumulate and cause mass destruction.

It is important to note that the immediate consequences of wildfires are not the only consequences that will continue to affect the area and the future likelihood of wildfires. There are many climatic and non-climatic factors that create a conducive environment for the start of wildfire spread. Some of these factors are direct climatic factors such as temperature, humidity, winds, and others are indirect climatic factors such as precipitation which will affect soil moisture and vegetation biomass. Other factors which are non-climatic include the topography, soil and vegetation type (Reiny, 2017). Many of these factors can be contributed to by the global warming of earth, causing an increase in annual mean temperatures, while other factors can be affected by the use of the local land areas and the way that the land is treated to yield optimal harvest or crops.

The Climate Change projections for Eastern Africa include increases in mean annual temperatures, bual distribution of the rainfall is expected to change, while the frequency and intensity of events such as droughts and floods is expected to increase and even possibly extend to new areas or locations. These projections for higher temperatures and the possibility for increased intensity of events combined with the likelihood for increased population growth pressure lends to a likely increase in occurrence for the intensity of wildfires in the region as well (Strydom & Savage, 2017).

Climate change combined with unsustainable management practices will increase wildfire risk by increasing the environmental vulnerability to wildfires. Therefore, it is increasingly important that there are sustainable practices in place to protect the environment and communities. Some such practices include the use of satellites and databases to monitor active fires, but more importantly helping to reduce climate change and protect our environment from these inclement changes.

Climate change has had a vast array of effects on the environment in Africa, many of which are contributors to wildfire and therefore will increase the likelihood for wildfire to spread to new areas. Examples include deadly floods in Nigeria, as well as droughts in Somalia since the start of 2022. The extreme weather events and disasters in Africa have been recorded to take the lives of at least 4000 people and affected another 19 million since the start of the 2022 year. Although, these numbers are likely to be much higher than those reported as the impacts of extreme events such as these often go unrecorded.

Africa is Not To Be Blamed

It is also important to note that although wildfires are a major consequence to the African peoples and the continental landmass, it is not just the African peoples who have contributed to their increase. The effects of climate change have had long term negative effects on the landscapes and environment as well as weather changes in Africa which thereby increase the risk for wildfire. This is despite the fact that Africa only contributes 4% of all global emissions, yet they are one of the most vulnerable in the world to climate change, according to the Intergovernmental Panel on Climate Change (IPCC). When comparing the effects of climate change on Africa when compared to other areas of the world, by using the Emergency Events Database (EM-DAT), the effect on Africa is equal to that of 40 individual extreme events (Dunne & Goodman, 2022). The EM-DAT is the largest database for disaster databases available, and it features extreme events. For an extreme event to be featured in this database, it must fulfill one of the following requirements: 10 more people dead, 100 or more people affected, the declaration of an emergency state, a call for international assistance. In 2022 alone, the death toll resulting from extreme weather events and disasters is 40x that (Dunne & Goodman, 2022).

Even before the end of 2022, the effects of climate change have shown to have major effects and negative consequences for some of the vulnerable populations across Africa. A well-known weather scientist at Imperial College London Dr. Friederike Otto stated that:

> "The biggest impact of climate change is not so much that single events have been made more extreme, but that there are even more extreme weather events in a region that already has always suffered from very high natural variability and high vulnerability. Just small changes in the number of extreme events are already having a huge impact (Dunne & Goodman, 2022)."

Dr. Otto makes an important point in his statement, which points out the incredible threat that climate change poses for these communities because even a small increase in extreme weather events in 2022 from prior years has already had such a large and widespread impact.

Effects of Wildfires on African Communities

In 2022 several areas in Northern and Central Africa experienced severe wildfires. For instance in February of 2022, severe forest fires were recorded in the Central African Republic amid the soaring temperatures and drought (Dunne & Goodman, 2022). Although there are many direct impacts of wildfires, these populations are hit hard by the secondary consequences of wildfires. According to a report by the UN OCHA, these fires in the Central African Republic forced over 500 families to seek emergency shelter and thereby led to disease outbreaks of malaria, cholera, and polio.

Similarly in July of 2022, Tunisia battled intense fires and high temperatures amid a head wave that exceeded 40 degrees celsius (Dunne & Goodman, 2022). These fires were not contained, but also broke out south of the capital city of Tunis, and once again led to the forced evacuation of several neighborhoods. The evacuation and emergency sheltering posed threats for disease outbreak, but were also the cause of damage to their grain crop.

Then again in August of 2022, Algeria also experienced severe wildfires as a result of the extreme heat. More than 100 fires were reported in North-East Algeria, killing at least 44 people and causing a further 2000 individuals to evacuate (Dunne & Goodman, 2022). These fires have affected thousands of people, including farmers who have lost dozens of hectares of land and hundreds of animals from their livestock. One such area in Souk Ahras lost one-third of its forests.

The factors behind every wildfire are complex, and although they will differ, research has shown that one of the largest contributing factors is climate change. With the changes in climate creating hot, dry conditions, "fire weather" is more common and perpetuates the increased risk of fires not only in Africa but also on a global scale. The most recent assessment done by the IPCC, projects that fire weather in northern, west southern, and east southern regions of the continent will only increase.

Dealing with Climate Change and Wildfires

Despite the current changes in climate and the major effects on the African subcontinent the current adaptation and mitigation interventions are not providing desired outcomes and instead are failing. As a result, transformative adaptation and mitigation measures are urgently needed to protect the land as well as the people and communities.

The United Nations is one organization that is currently pushing for the education and implementation of practices that will help communities live with wildfire. They are working on a "fire-wise" project that aims to strengthen disaster preparedness in the Fynbos Biome of South Africa which is a region that is extremely vulnerable to wildfires as a result of urbanization and agriculture. This area is one that has been fire-adapted and historically fire has been suppressed there, but this has resulted in a build-up of fire fields and thereby increased their vulnerability. The hope with this "fire-wise" project is implementation of adaptation measures that include information and communication technology that will anticipate the impacts of climate change (United Nations Climate Change, n.d.). The program would also teach people how to adapt to living with wildfire, and how to take action now to prevent future losses and will pilot practical adaptation approaches at the local level. This would include improving the capacity for local integrated fire management and developing a system that provides seasonal updates of fire fuel accumulation to calculate real-time fire danger risk.

Although there are organizations such as the United Nations and Greenpeace that work towards the establishment of fire prevention programmes, it is oftentimes difficult to fully implement. Prior to 2006, most African countries had already established a national fire prevention programme although the problem in most countries in sub-Saharan Africa is the lack of the basic resources or equipment to carry out these prevention strategies (Food and Agriculture Organization of the United

Nations, 2006). As well, the educational component is a key piece to implementation of successful programmes. Knowledge of basic fire behavior and the increase in efforts directed at education and training of local farmers will subsequently lead to a decrease in human caused wildfires but this still does not fix the issue entirely as most of these countries lack the capacities or resources.

With the rise in the number of disastrous events and extreme weather that has led to a rise in wildfires, the hope is that government priorities will change with the help of the United Nations and organizational support. For instance, the use of collaborative disaster management practices is necessary. In the United Republic of Tanzania, it has been suggested that there be joint involvement between the forestry staff and staff of the Fire and Rescue Service Force in order to pull resources and assist one another in disaster management and education to prevent widespread losses (Food and Agriculture Organization of the United Nations, 2006). Although this is a good use of resources, there are still other issues to resolve in creating successful prevention programmes such as leadership authority. For instance, who is in charge of conducting and implementing the fire suppression activities. These roles need to be clearly outlined and are especially important in commercial farming areas where farming practices are contributing to the fire risk. It is with great hope that the partnerships can be developed between institutions and governments along with outside organizations to provide the communities and leaders with resources and education necessary to carry out the successful implementation of wildfire reducing measures.

Conclusion

Climate change along with the slash and burn practices associated with farming and agriculture have resulted in the propensity for wildfires to occur in Africa. Although the majority of wildfires have been caused by humans in past years, the entire continent is starting to feel the effects of climate change caused by the carbon emissions of the rest of the world. Unfortunately, this leaves African communities to bear the burden of dealing with the consequences of the developing world when many of the hardest hit areas lack the resources, education, and equipment to do so. It is with great hope that the organizations involved are able to assist these communities in creating disaster management plans and equip these communities with the resources they need to prevent future losses although it is up to the entire global community to work at creating environmentally friendly measures to reduce the effects and speed at which climate change is affecting the world.

References

De Sales, F., Okin, G. S., Xue, Y., & Dintwe, K. (2019). *On the effects of wildfires on precipitation in Southern Africa.* Climate dynamics, 52(1), 951-967.

Dunne, D., & Goodman, J. (2022, October 26). *Analysis: Africa's unreported extreme weather in 2022 and climate change.* PreventionWeb. Retrieved December 19, 2022, from https://www.preventionweb.net/news/analysis-africas-unreported-extreme-weather-2022-and-climate-change

Food and Agriculture Organization of the United Nations. (2006). *Regional summaries.* Food and Agriculture Organization of the United Nations. Retrieved December 20, 2022, from https://www.fao.org/3/A0969E/A0969E03.pdf

Ichoku, C., Ellison, L. T., Willmot, K. E., Matsui, T., Dezfuli, A. K., Gatebe, C. K., ... & Habib, S. (2016). Biomass burning, land-cover change, and the hydrological cycle in Northern sub-Saharan Africa. *Environmental Research Letters,* 11(9), 095005.

Kayijamahe, E., & Otieno, V. (2020, June 15). *Wildfires in Eastern Africa — Will Climate Change Increase the Intensity of Wildfires?* ICPAC. Retrieved December 12, 2022, from https://icpac.medium.com/wildfires-in-eastern-africa-will-climate-change-increase-the-intensity-of-wildfires-573ba35a5e10

Reiny, S. (2017, January 9). *Study Finds a Connection Between Wildfires and Drought.* NASA. Retrieved December 10, 2022, from https://www.nasa.gov/feature/goddard/2017/nasa-study-finds-a-connection-between-wildfires-and-drought

Strydom, S., & Savage, M. J. (2017). *Potential impacts of climate change on wildfire dynamics in the midlands of KwaZulu-Natal, South Africa.* Climatic Change, 143(3), 385-397.

Styger, E., Rakotondramasy, H. M., Pfeffer, M. J., Fernandes, E. C., & Bates, D. M. (2007). *Influence of slash-and-burn farming practices on fallow succession and land degradation in the rainforest region of Madagascar.* Agriculture, Ecosystems & Environment, 119(3-4), 257-269.

United Nations Climate Change. (n.d.). *Reducing Wildfire Risks Associated With Anticipated Climate Change in the Fynbos Region – South Africa.* UNFCCC. Retrieved December 29, 2022, from https://unfccc.int/climate-action/momentum-for-change/activity-database/momentum-for-change-reducing-wildfire-risks-associated-with-anticipated-climate-change-in-the-fynbos-region

Wanzala, M. A., & Ogallo, L. (2020, June 5). *Recurring Floods in Eastern Africa amidst Projections of Frequent and Extreme Climatic Events for....* ICPAC. Retrieved December 17, 2022, from https://icpac.medium.com/recurring-floods-in-eastern-africa-amidst-projections-of-frequent-and-extreme-climatic-events-for-30d20d0d6f76

Chapter 4: Wildfires in Asia

Brianna Bedran

Introduction

In this chapter we will look at how exactly wildfires can start and what this means for the environment as well as the effect on one's health living in a region of recurring wildfires. A classification of different types of wildfires will also be examined in this chapter, and which ones are most common in Asia. Overall, we will find that petland and vegetation fires are familiar in regions throughout Asia. The smoke haze created by these fires has been reported to cause drastic negative ramifications on the air pollution in Asia, and this is further found to invoke cardiovascular and lung disease and infections. Moreover, the greenhouse gas emissions caused by the fires are having tremendous impacts on the environment of an already polluted region. To further demonstrate the significance of wildfires, we will look at the case of The Asian Forest Fires of 1997-1998 that exemplifies the issues we cover. We will deduce that some fires occur naturally, however there are large amounts that are bound by the economic pursuits of governments as well as climate change.

Causes of Wildfires in Asia

As a result of the increased trend of wildfires, near the end-of-the-century, regions throughout much of Southern Asia are estimated to emerge as critical and alarming wildfire hotspots that deserve our attention (Gannon et al., 2021). It is true that the persistent and intense haze events in Southeast Asia in the past decades have gathered the attention of not only governments but the public due to their impact on local economies, air quality and public health as their events have become more intense and frequent in recent years (Soejachmoen, 2019). However, due to the severity of the wildfires serious action must be taken along with this required attention. Studies reveal that there are a few factors that can lead to different kinds of wildfires in Southeast Asia. According to studies conducted by Hsiang-He Lee with the help of a team of researchers, burning biomass made up 40-60 percent of haze events in the major cities of Southeast Asia between 2003 and 2014 (Soejachmoen, 2019). As a result of the burning, the region, such as the mainland of Southeast Asia and the Maritime Continent, also faced a rise in carbonaceous compounds, including black carbon. Consequently, sunlight can be reduced through both absorption and scattering in the atmosphere, causing low visibility. Such events have been reported to hinder daily activities that are usually done effortlessly, including economic activities, air transportation and even students on their way to school (Soejachmoen, 2019). This information here allows us to understand the ways in which wildfires can create complexities in the everyday lives for civilians in Asia.

As for petland and vegetation fires, these are most often linked to burning activities for land use change purposes driven by economic motives (Phung et al., 2022). Economic motives for wildfires create sufficient damage and need to be looked at outside of the profitable pursuits of governments and instead focus on how much harm they are causing. These kinds of fires have been linked to the increasing smoke haze in the region which are raising alarming health and environmental concerns, matters which we will explore even further in the next section of this chapter to call attention to the harmful consequences. Furthermore, vegetation fires constitute a significant source of atmospheric trace gasses and aerosol particles (Langmann et al., 2009) and (Phung et al., 2022). Here the term vegetation fire signifies open fires of various vegetation such as savannah, forest, and agricultural residues, and petland fires that are set by humans or happen naturally, such as. by lightning (Langmann et al., 2009). There is an alarming and escalating global trend for the carbon emissions caused by vegetation fires between 1960 and 2000 going hand-in-hand with an increasing deforestation trend, in particular since 1970 in Southeast Asia (Langmann et al., 2009). Due to their escalating trend, the impacts of these fires are still paid for by the environment of Asia today. Additionally, conditions of dry weather caused by the El Niño-Southern Oscillation or a positive Indian Ocean Dipole event have been reported to create escalated fires in the region (Phung et al., 2022). Thus it is evident that there are many ways in which wildfires can be caused, which calls for the significance of understanding the impacts of wildfires and how to address them. While some can happen naturally, the economic pursuits of governments are preventable and given the severity of wildfires a reassessment of the value of these profits versus the harm of the environment and citizens is certainly called for..

Forest and petland fires in Indonesia have reportedly played the largest role in Asia's wildfires (Soejachmoen, 2019) which is why they are so often mentioned throughout this chapter. The fires themselves occur as a result of a combination of incidents such as those mentioned above, however, more intense long and hot dry seasons due to climate change can cause these fires (Soejachmoen, 2019). This is why it is difficult to address wildfires when they are caused by natural occurrences. Still, there are preventable actions of fires throughout Indonesia, such as fires that are quite often used to burn clear forests, agricultural residue, or prepare land for plantations and smallholder farms. Fire emission levels are greatest from degraded petlands, especially in dry years (Shannon N Koplitz et al., 2016). These occurrences are worse as most fires happen in peat areas, particularly in Sumatra and Kalimantan, which often spread rapidly from forests to the earth itself, creating more difficulties in extinguishing flames when the land and weather are dry.

Additionally, bigger fires during the last decades were made even worse by a drought brought on by the El Nino Southern Oscillation (ENSO). The delay of the monsoon in ENSO years results in fires that burn for several months longer than usual (Soejachmoen, 2019). This thick smoke that covered Equatorial Asia between the months of September and October in 2015 has been labeled as the worst haze episode Asia has seen with the exception of one in 1997, when land use fires caused billions of dollars in damage and thousands of premature deaths (Johnston et al, 2012) and (Marlier et al, 2013). In September–October 2015, El Niño and positive Indian Ocean Dipole conditions created the perfect recipe for tremendous fires in Sumatra and Kalimantan (Indonesian Borneo), establishing relentlessly hazardous levels of smoke pollution throughout much of Equatorial Asia. Hence, the 2015 smoke haze event was one of the most severe and prolonged transboundary air pollution events ever witnessed in Southeast Asia (SEA), altering the air quality of several countries within the region including Indonesia, Malaysia and Singapore (Sharma et al., 2015).

Vegetation and petland fires are attracting global attention due to their increasing frequency and magnitude of the fires. The events of such fires have been attributed to not only climate change, but as well as climatic and anthropogenic factors (Field et al., 2009) and (Phung, 2022). Vegetation fires consist of natural wildfires and prescribed fires for socioeconomic purposes (Phung, 2022). Meanwhile, petland fires consist of vegetation and the underlying peat layer, both of which are of critical concern in equatorial areas with large organic (histosol) and peat soil volumes. Natural climatic factors and prescribed fires are all crucial in playing a role in balancing ecosystem mechanisms and land management. However, the excessive amounts of unmanageable fires as a result of climate change have immense negative consequences on ecosystems and human health which will be navigated in the next sections of the chapter (Phung, 2022).

Wildfires Hotspots

For reference of where in South Asia call for the most concern of wildfires as of 2019, researchers have provided some numbers of wildfires hotspots. In South Asia, India had the highest number of annual fires followed by Pakistan and others whereas in Southeast Asia, Indonesia had the highest followed by Myanmar, Laos, etc. Frequency analysis has been found useful to identify hotspots of grid cells having recurrent fires (Vadrevu et al., 2019). Frequency analysis suggested nearly 30.5% of South/Southeast Asia with recurrent fires every year within a fifteen year time period with highest percentage in Laos (95.82%), Cambodia (78.7%), Thailand (75.0%), and Myanmar (75.1%) (Vadrevu et al., 2019). Becoming more aware and attentive of the wildfire hotspots in Asia is important in understanding the natural or human-caused occurrences of fires, which ones are preventable, and how to properly address them.

Impacts of Wildfires in Asia and Smoke Haze

In this section of the chapter, we will be looking into the impacts of wildfires in Asia and smoke haze, which have resulted in deterioration of air quality and lasting environmental consequences as well as health concerns for those residing in Asia. Municipalities throughout Southeast Asia are currently facing the consequences of both transboundary air pollution as well as their own local pollution, which results in urban haze (Soejachmoen, 2019). Studies have proven that out of the thirty-five most heavily polluted haze-event days in Singapore in October 2006, only seventeen days were associated with major outbreaks of burning in adjacent parts of Indonesia, and local pollution levels enhanced by stable meteorological conditions were the main reasons for poor air quality on the other 18 days (Soejachmoen, 2019). The trajectory analysis of researchers also signaled to the fact that pollution from Kalimantan in Indonesia had not reached Singapore during that haze period (Soejachmoen, 2019). Hence, smoke haze as well as an already polluted air quality in Indonesia are working together to create unfavorable air quality conditions in the region. Here we can acknowledge it is not just the smoke haze that creates poor air quality, but its impacts as well are exacerbated by the already polluted air. Air quality across Southeast Asia has been getting steadily worse for decades (Soejachmoen, 2019). The impacts of a lack of proper air quality in a region is not only difficult to manage, but also difficult to reverse given that once pollutants are emitted in the air there's no way to take it back. Researchers have confirmed that the deterioration is a combination of local pollution and pollution from upwind regions influenced by the dynamics of the atmosphere in the area, which is around the equator (Soejachmoen, 2019). The air quality index in Southeast Asia has been surpassing the unhealthy level, particularly during smoke haze periods, demonstrating how smoke haze due to vegetation and peatland fires in Southeast Asia is a serious public health concern that requires attention (Phung et al., 2022). The terms 'haze' and 'smoke haze' are used to describe extreme air pollution episodes due to burning activities on vegetation and peatlands (Behera et al., 2015) and (Latif et al., 2018). Particularly, haze most often refers to the high pollutant concentrations, as well as the low visibility conditions caused by the haze, and is generally used to represent extreme air pollution episodes not just limited to petland fires, but also for other sources from urban, industrial, and desert dust. Nonetheless, it is common to refer to smoke haze as a vegetation and petland fire-related air pollution episode in Southeast Asia (Phung et al., 2022)

While smoke haze is a regional issue in Southeast Asia, studies on its related health effects have been limited to only being reported from several countries in the region. The health burden of the smoke haze was found as excess mortality for all-causes, including but not limited to chronic respiratory diseases, lung cancer, cardiovascular diseases, and acute lower respiratory infection (Phung et al., 2022). Moreover, epidemiology (EPI) studies have also assessed large amounts of mortality and morbidity rates during smoke haze compared to non-smoke haze periods. Also demonstrating the harmful consequences of smoke haze, health burden assessment (HBE) studies estimated approximately 100,000 deaths attributable to smoke haze in the entire Southeast Asia (Phung et al., 2022). It is hard to imagine that simply partaking in one's daily mundane activities and doing something as seamlessly as exhaling oxygen is resulting in such excessive amounts of death in the region. However, this is unfortunately the reality and calls for more attention to be paid by government officials.

Wildfires and the Environment

Not only are the impacts of wildfires limited to critical health concerns, but also extends to the environment of Asia. The emissions of deforestation caused by wildfires in Asia are not compensated for any re-growth of necessary forestry in the region and have been providing a net source of CO_2 to the atmosphere and have severe impacts on the environment on top of the health consequences (Langmann et al., 2009). The consequences of wildfires on ecosystems can vary as a function depending on the severity and management of the fires (Vadrevu et al., 2019). As we are aware of by now, the management of wildfires plays a huge role in their repercussions, and the severity of wildfires can be demonstrated by a multitude of factors. There is some research that focuses on environmental concerns and some that may mostly focus on the health consequences, and in this chapter we have looked at both with the help of such information to demonstrate its significance in both forms. Further on the environmental influences, forest fires can have an impact in the loss of biodiversity which extends to the disruption of soil microbial processes during vegetation combustion and change important biogeochemical cycles. Fires have also resulted in landscape disturbance, which causes a combination of both burned and unburned forest patches, establishing a compound of heterogeneous patterns throughout the landscape (Vadrevu et al., 2019). The resulting landscape heterogeneity can further influence successional processes, which in turn may affect the spatial spread of subsequent fires (Vadrevu et al., 2019).. In addition, vegetation fires are known to release large amounts of greenhouse gas emissions including CO_2, CO, NOx, CH_4, non-methane hydrocarbons and other chemical species including aerosols impacting radiative budget, air quality and health at both local and regional scales (Vadrevu et al., 2019). To a greater extent, the pollutants caused by this can be transported over long distances impacting not only the local climate of Asia, but also the regional climate (Vadrevu et al., 2019).

It is evident that air pollution is not only a local issue in Asia as air has no boundaries. Thus, problems of air pollution can really only be resolved either by regional co-operation or global environmental laws, which have not yet been implemented (Soejachmoen, 2019). There is a critical and urgent necessity for innovative and effective transboundary international environmental legislations to check those nations that are the source of the pollutants from further deterioration. Concerned neighbors can resolve transboundary air pollution issues cooperatively and harmoniously. (Soejachmoen, 2019).

Considering all of these combined impacts also demonstrates the significance of characterizing fires in different regions of the world, not just including Asia (Vadrevu et al., 2019). In particular, quantifying vegetation fire trends and abnormalities help with identifying countries where fires have been either increasing or decreasing, and eventually relating the fire occurrences to driving factors such as climate, topography, vegetation and anthropogenic factors. An in-depth analysis of the type of vegetation burnt, fire intensities and also timing can help inform fire management at different spatial scales (Vadrevu et al., 2019). Researchers of wildfires signify the importance of these types of studies in identifying and preventing wildfires.

The Asian Forest Fires of 1997-1998

An incident of wildfires that displays both the severity and impacts of wildfires is The Asian Forest Fires of 1997-1998. The Asian Forest Fires provide more concrete examples of the issues we have explored in this chapter of environmental and health concerns. Further included in this section is economic impacts of fires as well using the 1997 case. During the months of October through November in 1997, Indonesia suffered from multiple fires and the consequential haze making them the headliners of media around the globe as the smoke haze surpassed as far as the Philippines to the north, Sri Lanka to the west, and northern Australia to the south (Mongabay, 2020). As a result, in Southeast Asia thousands of square miles of rainforest were destroyed by these fires, economies plummeted, and also burnt were conversion forests, plantations, and scrubland in over several areas of Indonesia (Mongabay, 2020).

Various nitrous oxides and sulfides, as well as ash were emitted in the air due to the burning. This combined with the industrial pollution and exhaust from cities, created choking haze that raised pollution levels to record-breaking heights (Mongabay, 2020). At this time, the air pollution Index of Indonesia transcended 800. This by far exceeds the normal amount of a typical day's exposure to an API which is on average 200-300 (Mongabay, 2020). Moreover, over 200,000 individuals were hospitalized with complaints of some less worrisome conditions of severe nosebleeds, and eye irritation, however many suffered ailments of heart and respiratory disorders (Mongabay, 2020). Moreover, there is also concern of the long-term health impacts for more than 70 million people in six countries who bore the haze. Health officials reported concern that the smog produced by the fires can potentially cause an increase in heart, lung, brain, eye, and skin disorders over the next decade (Mongabay, 2020).

The fires and haze exacerbated the faults of the region's economies, which were already in a bad state due to poor economic decisions. The fires were projected to have significant economic effects including a tremendous $3.15 billion loss from 1997 alone predicted by WAHLI, as well as the Economy and Environment Programme for Southeast Asia (EEPSEA) predicting that the total losses from 1997 and 1998 of roughly $5-6 billion after including factors such as a loss of plantations, biodiversity, timber, and for accommodating those with long-term health effects (Mongabay, 2020). Additionally, the smoke haze was so strong that it blocked out the sun in some regions, which diminished the ripening of fruits and prevented safe transportation (Mongabay, 2020). The effects of the lack of sun's influence on transportation extended to airport closings, boat accidents, and even a devastating Garuda Airbus a-300 plane crash that killed everybody on board. Tourism faced the consequences as well. Malaysia's second largest foreign-currency owner in 1996 at $4.5 billion, was wrecked by the haze, inspiring the government to issue a ban on all media coverage of the fires (Mongabay, 2020).

Conclusion

The repercussions of wildfires in Asia is an escalating problem that requires serious attention. Different classifications of fires are working together in the region causing a decrease in healthy levels of air quality and irreversible environmental impacts. Researchers of wildfires emphasize the necessity of researching, classifying, and responding to wildfires to do whatever can be done at this point to try and deescalate some of this damage. While there are wildfires that occur naturally, the profitable interests of governments that induce wildfires should be reassessed on whether these pursuits are really worth the damage. Given that the damage has created barriers and impacts in daily activities for people living in Asia, this is a considerable factor in any type of reassessment.

References

Behera SN, Betha R, Huang X, Balasubramanian R. (2015). Characterization and estimation of human airway deposition of size-resolved particulate-bound trace elements during a recent haze episode in Southeast Asia. Environ Sci Pollut Res. 2015;22: 4265–4280.

Environmental Defense Fund. (2022). "Here's how climate change affects wildfires". https://www.edf.org/climate/heres-how-climate-change-affects-wildfires

Field RD, van der Werf GR, Shen SSP. (1960). Human amplification of drought-induced biomass burning in Indonesia since 1960. Nat Geosci. 2009;2: 185–188.

Gannon, C. S., & Steinberg, N. C. (2021). A global assessment of wildfire potential under climate change utilizing keetch-byram drought index and land cover classifications. *Environmental Research Communications, 3*(3) doi:https://doi.org/10.1088/2515-7620/abd836

Hui Phung, V. L., Uttajug, A., Ueda, K., Yulianti, N., Mohd, T. L., & Naito, D. (2022). A scoping review on the health effects of smoke haze from vegetation and peatland fires in southeast asia: Issues with study approaches and interpretation. *PLoS One, 17*(9) doi:https://doi.org/10.1371/journal.pone.0274433

Langmann, B., Duncan, B., Textor, C., Trentmann, J., & van der Werf, G. R. (2009). Vegetation fire emissions and their impact on air pollution

and climate. *Atmospheric Environment*, *43*(1), 107–116. https://doi.
org/10.1016/j.atmosenv.2008.09.047

Latif MT, Othman M, Idris N, Juneng L, Abdullah AM, Hamzah WP, et
al. (2018). Impact of regional haze towards air quality in Malaysia: A
review. Atmos Environ. 2018;177: 28–44

Mongabay. (2020). "The Asian Forest Fires of 1997-1998". https://
rainforest.mongabay.com/
08indo_fires.htm

Sharma, R., & Balasubramanian, R. (2017). Indoor human exposure
to size-fractionated aerosols during the 2015 Southeast Asian smoke
haze and assessment of exposure mitigation strategies. *Environmental
Research Letters*, *12*(11), 114026. https://doi.org/10.1088/1748-9326/
aa86dd

Soejachmoen, Moetki. (2019). Tackling Southeast Asia's Air Pollution.
Global Asia. https://www.globalasia.org/v14no4/cover/tackling-
southeast-asias-air-pollution_moekti-h-soejachmoen
Vadrevu, K.P., Lasko, K., Giglio, L. *et al.* Trends in Vegetation fires in
South and Southeast Asian Countries. *Sci Rep* 9, 7422 (2019). https://doi.
org/10.1038/s41598-019-43940-x

Chapter 5: Wildfires in Australia

Diana Eve Amiscaray

Introduction

Australia is a beautiful island country known and celebrated for its gorgeous landscape and diverse wildlife. Browsing through travel catalogs would give you a glimpse of Australia's Great Barrier Reef, various beaches, national parks, and adorable animals such as koalas and kangaroos. A montage alone would illustrate the extent to which nature has become Australia's national treasure, however, the effects of climate change and wildfires may soon make the country unrecognizable.

Aside from Antarctica, Australia is the driest continent and an aerial view would reveal an image of its vast plains as a big desert (Britannica, n.d). In fact, the desert region takes up one-third of Australia's area, and a semi-desert part takes up another third. Only a limited portion of the country receives a sufficient amount of rainfall to allow for vegetation (Britannica, n.d).

Australia has already been characterized as a hot and dry country, and climate change is only making the situation worse. The impact of climate change has presented itself in Australia as blazing high temperatures, harsher droughts, and seasons of fire.

Wildfires, also known as bushfires, are a common phenomenon in Australia, which typically develop over the summer and affect dry inland areas the most (Cappucci, 2021). While this obliges officials to study wildfires and be equipped to handle emergency situations related to wildfires, nothing could have prepared Australia for the wildfires that occurred in 2019 and 2020.

The wildfires that emerged in Australia in 2019 and 2020 were like nothing anyone had ever seen before. This historic wildfire is estimated to have affected more than 42 million acres of land, and these flames were quick to become the source of lightning and smoke that travelled to the stratosphere (Cappucci, 2021). 2019 marked history as Australia's driest and hottest year and the fires that grew in this state left the country in dire need of manpower to extinguish the flames (Gunia, 2020).

Surprisingly, the overwhelming majority of people who respond to these emergencies are not professional firefighters; in New South Wales, about 90% of the heroes acting against the 2019-2020 wildfire crisis were ordinary Australian citizens who volunteered to help salvage their land (Gunia, 2020). Climate change has placed an enormous amount of pressure on Australia, and as the situation gets worse, the fires become more frequent and extreme. This has forced Australia to become dependent on its volunteers, many of whom have lost free time and income while dedicating themselves to the role of firefighters (Gunia, 2020). This should not have to be the reality for Australians, but is it possible to reverse the damage caused by wildfires? This chapter will focus on the impact of wildfires on Australia, and explore possible ways the country can progress toward a better environmental state.

The Impact of Wildfires on Australia

While the development of wildfires is certainly a national problem that affects all Australians, the states of New South Wales and Victoria have been particularly vulnerable (BBC, 2020). New South Wales has seen the worst of the situation, having its inhabitants displaced after over 2000 houses were destroyed by the bushfires in January 2020 (BBC, 2020).

Wildfires have affected Australia in many different ways. Two of the most visible consequences of the fires are related to the physical and ecological impacts on the country. As of January 2020, the United Nations Environment Programme estimates more than 18 million hectares of land were burned as a result of the 2019 to 2020 wildfire season, which caused the destruction of more than 2800 homes and more than 5900 buildings overall. The tragedy also cost the lives of millions of animals living in Australia (UNEP, 2020), including 25, 000 koalas in Kangaroo Island (BBC, 2020). Unfortunately, the effects of the wildfires persist even after the flames have been quenched. In regard to the fires' impact on biodiversity and the ecosystem, it is projected that around a billion animals and several more insects and bats are at risk of dying due to the loss of their habitats and food sources (UNEP, 2020). In a time when biodiversity is already taking a downturn, these wildfires prove to be a severe threat to Australia's ecosystem. Globally, terrestrial biodiversity is the richest in forest areas. In fact, more than 80% of all land animals, insects and plants reside in these forest areas (UNEP, 2020), which makes them particularly vulnerable to wildfire crises.

Wildfires are a huge issue for public health in Australia. Canberra, the capital of Australia, saw raging fires in the year of 2020 which prompted them to declare a state of emergency as the flames began approaching public places, including their airport and Parliament House (BBC, 2020). Reports from January 2020 revealed that out of all major cities worldwide, Canberra demonstrated the worst air quality index due to the large volume of smoke and air pollution induced by the wildfires (UNEP, 2020). The toxic smoke that results from wildfires can have fatal effects on humans because they contain particles and gasses that aggravate the respiratory system and eyes (UNEP, 2020). Furthermore, while the fires are able to spread to different neighboring areas, the resulting smoke has the ability to travel much farther than the flames. This harmful smoke gets pushed into the stratosphere by the heat emitted from the fires (UNEP, 2020). The World Meteorological Organization believes that the smoke from Australia's wildfires has been able to reach the Antarctic. In addition, the smoke has shown negative effects on the air quality of its neighboring country New Zealand, and remarkably, Argentina, Chile, and other cities in South America (UNEP, 2020). The fact that the smoke was able to travel to the other side of the world may suggest that Australia's wildfires may soon become an international problem.

Experiencing the effects of wildfires has taken a huge toll on the mental health and economy of Australia. The trauma acquired during the evacuation process alone is enough to trouble affected community members. Some Australians recount not being able to evacuate as soon as possible due to factors such as fuel stations not operating without electricity, or roads being blocked, preventing people from leaving areas deemed high-risk (UNEP, 2020). To make matters worse, the loss of shelter and possessions as a result of the fire is a nightmare that will continue to haunt affected Australians. While it is difficult to determine the monetary value associated with repairing the damage, it is estimated that the cost of wildfires is nearing 100 billion Australian dollars (Read & Denniss, 2020). AccuWeather CEO Joel N. Myers gives a higher estimate, calculating the economic loss to be 110 billion AUD considering the damage done to homes, businesses and their assets, infrastructure, agriculture, increased insurance premiums, power outages, firer fighting expenses and many other factors (Roach, 2020). Despite such high estimates from several economists, the Australian government responded with an emergency fund of 2 billion AUD (Quiggin, 2020). Of the 2 billion AUD emergency fund, 367 million AUD was proposed to be shared among the following sectors:

primary producers ($100 million), mental health ($76 million), local governments ($60 million), wildlife ($50 million), charities ($40 million), rural financial counselors ($15 million), financial counseling ($10 million), children's disaster payments ($8 million) and children's mental health ($8 million) (Karp, 2020). Following the bushfire season of 2019 to 2020, it has become clear that bushfires are the most costly natural disaster in Australia, and their destructive nature is a cruel reminder that climate change and its consequences can transcend generations.

Has Australia not learned from its past?

Australia's History of Wildfires

The occurrence of bushfires is not a recent issue. The records show that bushfires have affected Australia for centuries to the extent that historians even consider the country's landscape as "shaped by fire." This portion of the chapter will serve as a timeline of major bushfires throughout Australia's history. In addition, it will highlight the destructive aftermath of such bushfires and the conditions that led to them.

Black Sunday
Date: January 2, 1955
Location: South Australia

January 2, 1955 is an important date in Australian history. Referred to as "Black Sunday," this day was marked by horrific wildfires that plagued South Australia (Australian Disaster Resilience Knowledge Hub, n.d.). The city of Adelaide saw extreme heat measured at 43°C, paired with high-velocity gusts of wind, traveling at 100 kilometers per hour (Australian Disaster Resilience Knowledge Hub, n.d.) The end result of this tragedy was 40 thousand hectares of burnt land and two firefighters, killed in the bushfire (Australian Disaster Resilience Knowledge Hub, n.d.) A newspaper published at the time stated that "at least 50 people were injured and there were hundred more minor casualties" (The Canberra Times, 1955).

Black Friday
Date: January 13, 1939
Location: Victoria

Similar to the conditions described during the Black Sunday tragedy, forceful winds traveling through the state of Victoria facilitated the spread of fires on January 13, 1939 (Forest Fire Management Victoria, n.d.) This time around, it was recorded that the bushfires damaged about two million hectares of land throughout Victoria, destroying a large portion of the state's forest (Forest Fire Management Victoria, n.d.). Reflecting on the state of the environment and events happening at the time, it is thought that Black Friday was caused by both a period of drought and a combination of human actions. After a drought that endured for several years, Australia experienced a dry summer that dried up various bodies of water.

Black Tuesday
Date: February 7, 1967
Location: Tasmania

Black Tuesday refers to the bushfire season that occurred in South Eastern Tasmania in 1967. Notably, this tragedy started with 110 separate fires which quickly ended up transforming into several large flames (ABC, 2017). So far, a comparison between the events of Black Sunday, Black Friday and Black Tuesday reveals a common pattern: dry conditions, extreme heat, and rapid wind is a recipe for wildfires.

In 1996, the spring season brought an abundance of rain that allowed vegetation to flourish. However, after this period of prosperity came an unanticipated drought (ABC, 2017). Tasmania had not been as parched since the scorching summer of 1885 (ABC, 2017). By the time February came around, vegetation withered from the lack of precipitation. Paired with sweltering temperatures (39°C), powerful winds (100 kph), and low humidity, the dry conditions meant it was only a matter of time until Tasmania's inhabitants face the fire (ABC, 2017). What differentiates Black Tuesday from the previous two bushfires is perhaps the severity of the event. Black Tuesday resulted in 64 deaths, 900 injuries, tens of thousands of burnt hectares and 7000 people left without a home (ABC, 2017). Asides from the direct effect it had on those living in Tasmania at the time, it is estimated that the bushfire caused $40 million in damage (ABC, 2017).

1974-1975 Bushfires
Location: New South Wales

The bushfire season of 1974-1975 brought a huge amount of destruction to Australia. It has been recorded that 117 million hectares of land, which accounts for fifteen percent of Australia's physical land mass, was burnt from the wildfires (Chang, 2020). The death toll was much lower compared to that of other bushfire incidents, with only 3 lives lost as a result (Chang, 2020). Furthermore, the impact this bushfire season had on Australia can be estimated at 5 million dollars (Chang, 2020).

Ash Wednesday
Date: February 16, 1983
Location: Victoria and Southern Australia

Ash Wednesday occurred after a long period of dehydration. Prior to the development of fires, Australia suffered an extreme drought that lasted for ten months and ended up drying up the forests in eastern Australia

(Chang, 2020). Coupled with the dry conditions, a heatwave, strong winds and low humidity ignited the Ash Wednesday bushfire which claimed 75 lives (47 from Victoria and 28 from Southern Australia) and a total of 310, 000 hectares of land (160, 000 in Southern Australia and 150,000 in Victoria) (Chang, 2020). In addition, damage to infrastructure and livestock was extensive.

1993-1994

Black Christmas

Date: December 2001
Location: New South Wales (NSW) and Australian Capital Territory (ACT)

Fortunately, the Black Christmas bushfire season was not responsible for any deaths. On Christmas Eve (December 24), a series of over 100 fires affected communities located in New South Wales and Australian Capital Territory (Chang, 2020). In the 23 days of this bushfire season, the flames managed to burn 753, 314 hectares of land (Chang, 2020).

2003

Location: Canberra
The wildfires in Canberra claimed 4 lives, burned 160, 000 hectares of land, and left about 488 houses completely ruined (Chang, 2020).

Black Saturday

Date: February 7, 2009
Location: Victoria

Black Saturday can be considered the most tragic wildfire incident as it has resulted in the most deaths; 173 Australians lost their lives due to Black Saturday (Chang, 2020). Around 400 separate fires were involved in this disaster that left behind 450,000 hectares of burnt land, over 3000 homes and buildings wrecked, and over 11,800 livestock lost (Chang, 2020).

Causes of Bushfires and Factors that Influence their Development

As demonstrated in the previous few sections of this chapter, bushfires can occur in a considerably spontaneous manner. A combination of different environmental factors make the development of bushfires probable, however, this fact does not negate society's responsibility in the prevention and mitigation of bushfires.

Firstly, climate and landscape are huge factors in determining how fire-prone a region is. Almost all types of vegetation in Australia are prone to fire due to the country's geographic location and topography (Australian Bureau of Statistics, 1995). The only areas that are theoretically spared from the fire are the tropical rainforests situated in north Queensland (Australian Bureau of Statistics, 1995). In general, there are certain months associated with wildfires, however, fire season actually varies across different parts of Australia. Specifically, fire season is dependent on latitude (Australian Bureau of Statistics, 1995).

Fire Seasons
Northern Australia: Winter and Spring
Southern Australia: Summer and Autumn

Regions with tall forests are considered to be rich in fuel, and thus, flammable. In the event that these forests dry out, such abundance in fuel can give rise to severe flames (Australian Bureau of Statistics, 1995).

Interestingly, there are people who do not blame climate change for the increasing incidence of wildfires, rather, they point their fingers at fellow Australians. Surprisingly, several fires contributing to the growth of bushfires were deliberately lit. In fact, going back to Black Tuesday, it is thought that out of 110 fires, 88 were intentionally ignited

(ABC, 2017). In fact, in order to promote conscientious behavior from Australians, the government has ensured that negligent actions face legal consequences. In the recent 2019-2020 bushfire season, police took legal action against 159 people who were thought to contribute to fires (Cole, 2020). Moreover, if found responsible for starting a wildfire, an Australian can be sentenced to a maximum of 21 years in prison (Cole, 2020).

Fire Management Practices of Indigenous Australians

Australia is home to a diverse population of people who identify with different cultures, languages and heritages. Among the country's inhabitants are the Indigenous peoples of Australia, grouped into two categories: Aboriginal Australians and the Torres Strait Islander peoples. As the first to live in Australia, the Indigenous peoples possess a profound wisdom and deep understanding of the country's landscape. In fact, it is thought that Aboriginal Australians were very knowledgeable about fire management. While this might seem counterintuitive, Aboriginal Australians seemed to protect their land from wildfires by starting fires themselves. Fires were a central part of the Aboriginal routine since they were used for hunting, warmth, protection against snakes, warfare and several other activities (Australian Bureau of Statistics, 1995). Despite the frequent use of fire in the Aboriginal lifestyle, fires were not as intense even if the weather permitted the growth of extreme flames (Australian Bureau of Statistics, 1995). This phenomenon can be explained in principle that the fuel load was low due to frequent burning (Australian Bureau of Statistics, 1995). Furthermore, it seems that some Aboriginal Australians would start a fire for the sole purpose of cleansing the country (Australian Bureau of Statistics, 1995).

The Future of Australia

Reflecting on the timeline of historic bushfires, traditional fire management practices, and the most recent 2019-2020 bushfire season, how should Australians proceed? What is the best course of action? Can we counteract the effects of climate change?

Record of historic bushfires and the details of their events teaches Australians what conditions favor the development of wildfires. As illustrated in the timeline, many of the notable events occurred in January or February and were preceded by conditions of extreme drought. Since the country has a comprehensive history of wildfires, Australians can be familiar with combustible conditions and may be able to act proactively. Investment in fire management services can be useful, especially in fire-prone areas and before anticipated seasons. Research is already being conducted into methods such as prescribed burning, which is quite similar to the practices of Indigenous Australians before European settlement. However, some studies such as that of Clarke et al. suggest that prescribed burning alone is insufficient in the prevention of wildfires (Clarke et al., 2022). The study reveals that recent wildfires, such as those in the 2019-2020 season, are more severe compared to historic ones. Furthermore, while prescribed burning lowers the residual risk of wildfires, it would take an excessive amount of treatment to match the residual risk associated with historic wildfires, or even below that value (Clarke et al, 2022). In other words, the study suggests that there is no going back. Wildfires have gotten more aggressive over the years and this trend will likely continue. Prescribed burning may lower the potential severity of future wildfires, yet it would still not compare to the relative mildness of early wildfires. This is largely due to climate change, which is a global issue that continues to worsen. Climate change promotes conditions that favor wildfires, thus, it is imperative to adopt attitudes that prevent its exacerbation.

Conclusion

Overall, wildfires or "bushfires" are a common phenomenon in Australia that helped shape its landscape. However, as climate change becomes a bigger problem globally, Australia becomes more vulnerable to wildfires. The aftermath of bushfires is devastating; families are forced to evacuate their homes, victims are faced with property loss, businesses lose income, millions of animals lose their habitats, and the country as a whole is left in grief. The recent 2019-2020 bushfire season was the worst Australia had ever seen up to this point, and there are huge concerns regarding whether or not the nation is equipped to manage similar fires in the future. While it is hard to foresee what the future holds, one thing is for sure: the dedication of Australians to their community and protecting their land makes for a powerful force against the great burn.

References

ABC News. (2017, February 6). *Loss and lessons from Tasmania's Black Tuesday bushfires*. ABC News. Retrieved December 31, 2022, from https://www.abc.net.au/news/2017-02-06/tasmanias-1967-black-tuesday-bushfires-explained/8241698

Australian Bureau of Statistics. (1995, January 1). *Bushfires - an integral part of Australia's environment (feature article)*. Australian Bureau of Statistics. Retrieved December 30, 2022, from https://www.abs.gov.au/Ausstats/abs@.nsf/0/6C98BB75496A5AD1CA2569DE00267E48

Australian Government National Emergency Management Agency. (n.d.). *Bushfire-New South Wales* . Australian Disaster Resilience Knowledge Hub . Retrieved December 30, 2022, from https://knowledge.aidr.org.au/resources/bushfire-new-south-wales-1974/

BBC. (2020, January 31). *Australia fires: A visual guide to the bushfire crisis*. BBC News. Retrieved December 17, 2022, from https://www.bbc.com/news/world-australia-50951043

Bushfire - black sunday, Mt. Lofty Ranges, 1955: Australian disaster resilience knowledge hub. Black Sunday, Mt. Lofty Ranges, 1955 | Australian Disasters. (n.d.). Retrieved December 17, 2022, from https://knowledge.aidr.org.au/resources/bushfire-black-sunday-mt-lofty-ranges-1955/

Cable News Network. (2020, January 10). *Australia is promising $2 billion for the fires. I estimate recovery will cost $100 billion | CNN business.* CNN. Retrieved December 17, 2022, from https://edition.cnn.com/2020/01/10/perspectives/australia-fires-cost/index.html

Cappucci, M. (2021, July 27). *Australian fires had bigger impact on climate than covid-19 lockdowns in 2020.* The Washington Post. Retrieved December 17, 2022, from https://www.washingtonpost.com/weather/2021/07/27/australian-bushfires-smoke-climate-covid/

Chang, C. (2020, January 10). *How this year's fires compare to others.* NewsComAu. Retrieved December 30, 2022, from https://web.archive.org/web/20200204103523/https://www.news.com.au/technology/environment/how-the-2019-australian-bushfire-season-compares-to-other-fire-disasters/news-story/7924ce9c58b5d2f435d0ed73ffe34174

Clarke, H., Cirulis, B., Penman, T., Price, O., Matthias, M.B & Bradstock, R. (2022). The 2019-2020 Australian Forest Fires are a Harbinger of Decreased Prescribed Burning Effectivenees under Rising Extreme Conditions, *Nature*, 12(11871). https://doi.org/10.1038/s41598-022-15262-y

Cole, B. (2020, January 7). *Amid worst australia wildfires for a decade, 24 people charged with arson.* Newsweek. Retrieved December 30, 2022, from https://www.newsweek.com/australia-wildfires-arson-new-south-wales-police-1480733

Encyclopædia Britannica, inc. (n.d.). *Land of Australia*. Encyclopædia Britannica. Retrieved December 17, 2022, from https://www.britannica.com/place/Australia/Land

Forest Fire Management Victoria. (2021, July 2). *Black Friday 1939*. Forest Fire Management Victoria. Retrieved December 17, 2022, from https://www.ffm.vic.gov.au/history-and-incidents/black-friday-1939

Guardian News and Media. (2020, January 17). *Bushfire recovery: How is Australia's $2bn fund being spent?* The Guardian. Retrieved December 17, 2022, from https://www.theguardian.com/australia-news/2020/jan/18/bushfire-recovery-how-is-australias-2bn-fund-being-spent

Gunia, A. (2020, December 11). *Australia's firefighters: Time's heroes of the year 2020*. Time. Retrieved December 17, 2022, from https://time.com/collection/heroes-of-2020/5916451/australias-firefighters/

Read , P., & Denniss , R. (2022, August 4). *With costs approaching $100 billion, the fires are Australia's costliest natural disaster*. The Conversation. Retrieved December 17, 2022, from https://theconversation.com/with-costs-approaching-100-billion-the-fires-are-australias-costliest-natural-disaster-129433#:~:text=The%20Deloitte%20Access%20Economics%20ratio,of%20Queensland%20economist%20John%20Quiggin.

Roach, J. (2020, January 8). *Australia wildfire damages and losses to exceed $100 billion, AccuWeather estimates*. AccuWeather. Retrieved December 17, 2022, from https://www.accuweather.com/en/business/australia-wildfire-economic-damages-and-losses-to-reach-110-billion/657235

UN Environment Programme. (2020, January 22). *Ten impacts of the Australian bushfires*. UNEP. Retrieved December 17, 2022, from https://www.unep.org/news-and-stories/story/ten-impacts-australian-bushfires

Wide damage in South Australia - Adelaide, Monday. - The Canberra Times (act : 1926 - 1995) - 4 jan 1955. Trove. (n.d.). Retrieved December 17, 2022, from https://trove.nla.gov.au/newspaper/article/91201308

Chapter 6: Wildfires in the Mediterranean

Amal Rizvi

A Brief Introduction to Forest Fires

The Mediterranean regions are a very broad area across the entire globe; in fact, Mediterranean climates span across the five continents of South America, North America, Africa, Australia, and Europe (Moreira et al., n.d.). However, for the sake of simplicity, this chapter will focus on the regions in Europe which surround the Mediterranean basin. These areas include Italy, Greece, Spain, and France (Moreira et al., n.d.).

Mediterranean regions have some very distinct qualities that contribute to the ever increasing presence of forest fires in the area. First of all, they experience a seasonal climate in which the summer months are extremely dry and warm, whereas the months of the winter season are wet and cold (Moreira et al., n.d.). Thus, as one would likely expect, winter months in the Mediterranean allow for the growth of many species of vegetation. Vegetation in the context of forest fires can be thought of as fuel for fires, because it is often readily flammable. After the growth of vegetation in the winter months, the extremely dry, hot summers facilitate the ignition of fuel. As one might predict, when the climate warms due to global warming, this results in the increase of forest fires in the Mediterranean.

Fires in the Mediterranean Regions: How are they linked to humans and what are the long-term effects?

Forest fires are rather complex, because they are the product of both natural, as well as anthropogenic (meaning, the result of human intervention) causes (San-Miguel-Ayanz et al., 2013). For over tens of thousands of years, humans living in and around the Mediterranean have been using fires in their day to day lives (Daniau et al., 2010). In fact, using fire is still a reliable management tool in several parts of the Mediterranean regions; it is a great option when it comes to controlling the growth of vegetation in rural areas and creating pasture that is used to feed cattle (San-Miguel-Ayanz et al., 2013). That being said, the vast majority of fires that occur in the Mediterranean today are actually not the product of human necessity, but rather unnecessary human intervention (San-Miguel-Ayanz et al., 2013). Fire use for land management has been mostly phased out as human populations have stopped occupying many areas within the Mediterranean where the land is simply not productive enough to support agriculture (Pausas and Fernández-Muñoz, 2012).

That being said, it is important to make a distinction: while all forest fires in the Mediterranean region are not directly the cause of human impact, much of the forest fires that occur throughout the Mediterranean countries are very much intertwined with humans. Why is this, one may ask? Well, first of all, forest fires have very large social, economic, and environmental costs, especially throughout countries like Spain, Italy, Portugal, and Greece (World Wildlife Fund [WWF], n.d.). Secondly, ecosystems in the Mediterranean region are losing their ability to naturally regenerate; while some fires are indeed the consequence of natural ecosystem dynamics and allow for forested areas to regenerate stronger, fires that occur too frequently have the opposite effect (WWF, n.d.). They threaten the stability of ecosystems and result in the loss of biodiversity, as well as soil erosion and reduced water in the area (WWF, n.d.). Fires in the Mediterranean regions also result in changed climate. As of right now, the climate in the Mediterranean region is shifting towards long, intense droughts in the summertime and more extreme weather events (WWF, n.d.).

Currently, roughly 60,000 fires occur in the Mediterranean region every single year (San-Miguel-Ayanz, 2009). These fires burn approximately half a million hectares of forest area in the region (San-Miguel-Ayanz, 2009). It is difficult to obtain an exact number of these fires, because statistics only began being collected in the 1970s (San-Miguel-Ayanz, 2009). Much of this data is contained in the European Forest Fire Information System, also known more concisely as EFFIS (San-Miguel-Ayanz, 2009).

History of Fires in the Mediterranean Region

Fires have a complicated history in the Mediterranean. As mentioned previously, they have been used historically as a management tool by humans for a long time. Paleolithic people created intentional fires in order to help them with their hunt and food gathering (Rackham, 2003). However, it is said that changes to the Mediterranean landscape as a result of fires were first noted in the Neolithic time period. Other evidence of fire in these times has been identified by remnants of charcoal in pollen deposits (Rackham, 2003). However, this piece of evidence is not fully conclusive due to charcoal deposits being rare. Also, the deposits do not easily identify how much land was burnt, or at what time of year (Rackham, 2003).

Evidence from the Middle Ages and onwards suggests that new rules surrounding intentional occupational burns began to arise (Rackham, 2003). Accounts from travelers after this period of time also outline the use of fire (Rackham, 2003). One of the major regulations that came about was a decree by the Senate of Venice in 1414 (Rackham, 2003). The decree was the result of people noticing that cypress, a tree that was sensitive to fire, was in decline in the region (Rackham, 2003). This is an early recount of people noticing biodiversity decline and other harms as a result of fires. Another example of this is documented in the Carta de Logu, which is the 14th century law code of Sardinia (Rackham, 2003). In this code, it was outlined that wildfires must be kept far away from the town (Rackham, 2003).

Since the 1900s, the amount of surface area burnt has increased exponentially; between 1960 and 1973, only about 50 kilo hectares of land was burnt and less than 2000 fires were noted (Pausas and Vallejo, 1999). This number continues to increase to this day.

Causes of Fires in the Mediterranean Regions: Land Use Change

There are multiple reasons why fires continue to occur in the Mediterranean region, especially with the increased frequency being seen in recent years. The first and major reason is land use change as a result of human intervention (Pausas and Vellejo, 1999). These days, the use of fire management to manipulate vegetation is not common, but humans continue to intervene with the Mediterranean land in other ways to develop their infrastructure (Pausas and Vallejo, 1999). Europe has undergone extensive industrial development which has resulted in people moving away from rural areas and into urban settings, like cities (Pausas and Vallejo, 1999). Further, Europe has seen a major shift from grazing pressure and wood gathering to agricultural mechanization (Pausas and Vallejo, 1999). What is the net result of this, one may ask? Well, people are continuously abandoning rural land, which allows for vegetation to grow.

While vegetation growth may seem like a positive thing, one must realize that when humans abandon large areas of land, fuel accumulates in those areas. It is this fuel that gives rise to forest fires (Pausas and Vallejo, 1999). Tourist pressure is another way fuel can accumulate. This is because increasing demand for tourism once again results in people choosing to conglomerate in specialized tourist areas, which are typically around urban centers. Similarly, this promotes people moving away from rural landscapes and facilitating fuel accumulation.

Evidence of this is noted in forest fire trends in the Mediterranean. In the southern areas of the Mediterranean region, people still adhere to traditional land use, and there is not as great of a shift towards urbanization; as a result, forest fires are less frequent due to less accumulation of fuel (Pausas and Vallejo, 1999).

Causes of Fires in the Mediterranean Regions: Climate Change

The second major cause of forest fires occurring at an increasing rate in the Mediterranean regions of the world is climate change (Pausas and Vallejo, 1999). What is climate change, exactly? Well, climate change is the incessant warming of Earth's surface temperature over the long term. Much of climate change is the consequence of the anthropogenic greenhouse effect, which is the gradual, but incessant trapping of heat due to gasses humans have released into the Earth' atmosphere - namely carbon dioxide. While Earth has a natural greenhouse effect as well, the effects of the anthropogenic addition of greenhouse gasses has accelerated climate warming at an alarming rate.

How do warming temperatures on Earth result in forest fires in the Mediterranean region? Well, in this case it is important to remember that forest fires in the Mediterranean regions occur in the hot, dry summer months; the air is not very humid at all during this time, and fuel is low in moisture (Pausas and Vallejo, 1999). Plants are becoming increasingly water stressed as the temperature increases as well (Pausas and Vallejo, 1999). All of these contribute to the risk of forest fires in the Mediterranean increasing, as the probability of ignition increases (Pauses and Vallejo, 1999).

Major Mediterranean Wildfires in Recent Years

The previous section mentioned a brief history of fires occurring in the Mediterranean region since prehistoric times. However, wildfires are common in this area of the world even today. In fact, the Mediterranean region is known as a hotspot for wildfires. The next few sections will outline major fires that occurred in the Mediterranean region in recent years.

#1. The 2022 European and Mediterranean Wildfires

Intense wildfires are still occurring to this day in the Mediterranean region. In 2022, roughly 147,900 hectares of land, or 365,000 acres, was burned (Hogger, 2022). The burns began in July 2022. With record-breaking heat waves occurring in much of Europe this past summer, drought conditions arose shortly after and began giving rise to wildfires. Some of the most significantly impacted areas were Spain, France, Portugal, Greece, and Turkey. The fires caused mass evacuation of communities across the Mediterranean, the closure of major roads, and other disturbances. The 2022 wildfire period across the Mediterranean set records; in Spain, it was the worst record of a wildfire since 2003, for instance.

#2. The 2009 Mediterranean wildfires

In 2009, a devastating series of wildfires occurred across several Mediterranean countries (McVeigh, 2009). These countries included

Greece, Italy, France, Spain, and Turkey. During this time period, the temperatures were especially hot and dry during the summer, with temperatures rising as high as forty-four degrees Celsius.

When it comes to reasons these wildfires occurred, it is important to note a few things. Firstly, farmers in this region at the time were partaking in great amounts of uncontrolled legal and illegal scrub burning (McVeigh, 2009). Scrub vegetation in the Mediterranean mainly consists of shrubs that are adapted to fire (Plant Life, 2011). This is because, as mentioned previously, the climate in this region is dry in the summertime, and natural fires in the region are common. Thus, the shrubs have, over time, developed adaptations, such as underground structures that remain intact even when fires pass over the land (Plant Life, 2011). These structures allow for new growth after the fire is over. Other adaptations include long-lived seeds that actually require heat to induce germination.

How has Policy Influenced the Increasing Fires in the Mediterranean Regions?

Policy is a major way in which humans can manage large scale events, such as forest fires. While the fire itself may be out of people's control, good policies and correct management can help minimize destruction, economic and social consequences, as well as fire frequency over time. Currently, existing policies surrounding the fires in the Mediterranean are extremely inadequate (Moreira et al., n.d.). One may wonder, "What is the reason for this?". Well, the policies in the Mediterranean region do not account for the two major causes of the fires that were previously mentioned – land use change and climate change. This has resulted in something called a "firefighting trap" (Moreira et al., n.d.).

What in the world is a firefighting trap? In short, it is a process by which people incorrectly assume that the best way to handle frequent fires is to suppress symptoms, rather than actually eliminating the true underlying reason for which these fires occur (Xanthopoulos et al., 2020). The end result of this is a higher probability that forest fires will occur in the future at an increased frequency. Firefighters in the Mediterranean regions tend to use fire suppression, which is a shortsighted approach because it operates with the end goal of minimizing area burned. Areas burned by fires, although a rather devastating consequence, is a symptom of a larger issue, but not an adequate way to correctly minimize fire frequency. Further, when these large-scale fires occur in the Mediterranean region, the response to public outcry and opinion is often governments investing even more money in firefighting – thus accelerating the firefighting trap – rather than greater research into reduced fire hazard and risk.

Alternative Options

In order to control the ever increasing fires in the Mediterranean region, it is suggested that countries in this area move away from expending energy, time, and money into firefighting efforts, and instead aim to minimize the damage, rather than the area that is burned (Moreira et al., 2020). This is because as climate warming continues to impact the Mediterranean region, this part of the world is only getting warmer, rainier, and more productive with time. A more productive region means that biomass is steadily increasing, and more flammable material is becoming available. Thus, if firefighting efforts continue and governments focus on minimizing area burned, the fires in the Mediterranean region are only delayed, not eliminated.

Thus, the other negative impacts – namely, the damage – that are caused by fires should be attended to instead (Moreira et al., 2020). Policies must also be changed to reflect this. Some of the areas policy makers should be focusing on, rather than firefighting efforts, are number of human lives affected and/or lost, economic consequences, and the quality of the air, water, and soil after a wildfire takes place.

Current Recommendations to Manage Wildfires in the Mediterranean Region

As of right now, there are several suggestions that management groups have put out in order to better control the forest fire situation in the Mediterranean region (Food and Agricultural Organization of the United Nations [FAO], n.d.). The first suggestion has to do with involving large groups with the power to influence public policy on fire management (FAO, n.d.). Such groups include the Committee on Forestry. Further, it is encouraged to use events as a main source of drawing attention on the prevention of wildfire. For instance, an event such as Mediterranean Forest Week could be a key time in which speakers and policy makers can come together to share and promote the latest information on forest fires in the Mediterranean, as well as preventative measures. These events not only serve to bring powerful groups and people together, but they can also involve the public in decisions. Overall, it is evident that multiple groups and sectors must cooperate to tackle the mutli-faceted issue of forest fires.

Once the right groups and people have come together, the next major step is to actually change public policies on forest fire management that take into account crucial factors, such as climate warming and land use changes over time (FAO, n.d.). These policies must be thorough; they may involve incentives and/or obligations that concern preventative actions by municipalities, land owners, and building enterprises. In these policies, the roles of forest also must be examined and re-structured; even today, many people are not aware of the larger implications healthy forests can have on human lives. If the prevention of forest fires is promoted as a very much necessary step in forest management, and people understand the importance of forests as a whole, then people may be more likely to agree with policy changes.

Conclusion

The information presented in this chapter, as well as the vast number of news reports on forest fires that occur every single year in the Mediterranean make one thing very clear: fires in this region of the world are an extremely prevalent issue that can not, under any circumstances, be ignored. While fires may indeed be considered a natural occurrence in healthy ecosystems, these specific fires in the Mediterranean are not entirely natural. Human influence has resulted in the frequency and intensity of fires to drastically increase at an alarming rate. While it is true that many old age practices that involve fire management, such as using fire for the gathering of crops, are not as common, human presence in the Mediterranean has a dramatic effect on the land today.

Land use change in conjunction with climate change are the two major driving forces that have amplified forest fires in this part of the world. Both of these causes are largely the result of humans. In the case of land use change, humans are leaving rural areas behind in favor of urban centers like cities. This mass movement has cascading effects on fuel accumulation and burning in the Mediterranean forests. Tourism also greatly impacts this. In terms of climate change, while some climate change is expected and very much natural, human induced climate change as a result of excessive greenhouse gas emissions (I.e. fossil fuel burning) has resulted in hotter, drier Mediterranean summers. It is these warmer temperatures that give rise to forest fires because of an increased probability of ignition.

In order to tackle the issue of forest fires in this region, policy shifts are required. More specifically, policies must be changed from focussing on firefighting to decreasing the other negative impacts that fires cause. Aside from area burnt, impacts such as the number of human lives affected, as well as the socioeconomic and cultural effects of frequent forest fires need to be examined in a broader context.

References

Daniau, A.-L., d'Errico, F., & Sánchez Goñi, M. F. (2010). Testing the hypothesis of fire use for ecosystem management by Neanderthal and Upper Palaeolithic modern human populations. PLoS ONE, 5(2). https://doi.org/10.1371/journal.pone.0009157

Hogger, C. (2022). Southern Europe wildfire outlook – summer 2022. Crisis24. Retrieved January 5, 2023, from https://crisis24.garda.com/insights-intelligence/insights/articles/southern-europe-wildfire-outlook-summer-2022

McVeigh, T. (2009, July 25). Hundreds evacuated as med coast wildfires spread. The Guardian. Retrieved January 5, 2023, from https://www.theguardian.com/world/2009/jul/26/summer-wildfires-spain-mediterranean-coast

Mediterranean scrub. Plant Life. (2011). Retrieved January 5, 2023, from http://lifeofplant.blogspot.com/2011/03/mediterranean-scrub.html

Moreira, F., Ascoli, D., Safford, H., Adams, M. A., Moreno, J. M., Pereira, J. M., Catry, F. X., Armesto, J., Bond, W., González, M. E., Curt, T., Koutsias, N., McCaw, L., Price, O., Pausas, J. G., Rigolot, E., Stephens, S., Tavsanoglu, C., Vallejo, V. R., … Fernandes, P. M. (2020). Wildfire management in Mediterranean-type regions: Paradigm change needed. Environmental Research Letters, 15(1), 011001. https://doi.org/10.1088/1748-9326/ab541e

Pausas, J. G., & Fernández-Muñoz, S. (2011). Fire regime changes in the western Mediterranean Basin: From fuel-limited to drought-driven fire regime. Climatic Change, 110(1-2), 215–226. https://doi.org/10.1007/s10584-011-0060-6

Pausas, J., & Vallejo, V. (1999). Remote sensing of large wildfires in the European Mediterranean Basin. Springer.

Rackham, O. (2003). Fire in the European Mediterranean. University of Arizona. Retrieved January 5, 2023, from https://ag.arizona.edu/OALS/ALN/aln54/rackham.html

San-Miguel-Ayanz, J., Moreno, J. M., & Camia, A. (2013). Analysis of large fires in European Mediterranean Landscapes: Lessons learned and perspectives. Forest Ecology and Management, 294, 11–22. https://doi.org/10.1016/j.foreco.2012.10.050

San-Miguel-Ayanz, J., Pereira, J. M. C., Boca, R., Strobl, P., Kucera, J., & Pekkarinen, A. (2009). Earth observation of wildland fires in Mediterranean ecosystems. Scholars Portal.

World Wildlife Fund [WWF]. (n.d.). Forest fires in the Mediterranean: A burning issue. European Union. Retrieved January 5, 2023, from https://environment.ec.europa.eu/topics/forests_en

Xanthopoulos, G., Leone, V., & Delogu, G. M. (2020). The suppression model fragilities. Extreme Wildfire Events and Disasters, 135–153. https://doi.org/10.1016/b978-0-12-815721-3.00007-2

Chapter 7: The Health Effects of Wildfire Smoke

Daniel Klassen

Introduction

The New Democratic Party of Alberta, in preparation for the 2023 Alberta Election, has made promises to update legislation to include presumptive Workers' Compensation coverage for firefighters who responded to the 2016 Fort McMurray wildfire. The fire was so catastrophic it gained the moniker of "the Beast" after destroying hundreds of homes and businesses within Fort McMurray, Anzac, and surrounding areas within the Regional Municipality of Wood Buffalo. (Parrish, 2018) Firefighters battling the wildfire have argued that current coverage rules left many ineligible for cancers. Breathing problems have also arisen in recent years. The proposed adjustment to legislation adds respiratory illnesses and cancers, including thyroid and pancreatic cancers, to the current list of presumptive firefighter cancers. The proposal also removes the latency period on these cancers for all firefighters who responded to the wildfire. (McDermott, 2022) The New Democrat Party's plan to adjust Worker Compensation coverage generates several existential questions; What is the danger of wildfire smoke? How does wildfire smoke affect the general public? What can individuals do to keep their families and communities safe?

The Toxic Twins

Every firefighter is instructed on the dangers of the so-called "Toxic Twins," two chemicals that have synergistic effects. During the burning of combustible materials, carbon monoxide and hydrogen cyanide create a deadly chemical asphyxiant. Smoke inhalation is complicated, even for firefighters. One target organ of hydrogen cyanide is the heart. Most structural firefighters have access to self-contained breathing apparatus (SCBA), the primary component of personal protective equipment (PPE) to keep firefighters from being exposed to toxic smoke and its subsequent effects on human health. (Burke, 2016) Similar to firefighters, there are benefits for the general public to educate themselves on smoke inhalation. Determining a diagnosis of cyanide poisoning is challenging because, unlike carbon monoxide, there is no readily available test for the presence of cyanide within victims. (Holstege et al., 2015) Comprehending that cyanide poisoning is a possible outcome of smoke inhalation could lead to more health-sustaining awareness within the general public.

Many people worldwide can recall the smell of a campfire, a wood-burning fireplace, or the sulfurous smell following the strike of a match. Unfortunately, for some communities, the smell of smoke may trigger memories of being enveloped in wildfire smog unabated for several weeks. The California wildfire season was catastrophic, with over 4.3 million acres burnt, over ten thousand structures lost, and 33 fatalities. (CalFire, 2022) Various forms of pollution from vehicles, industry, pesticides, fertilizers, and waste may enter the environment, contaminating soil and plants. When organic materials burn, toxic chemicals are released along with particulate matter in the smoke, with burning buildings or infrastructure adding more chemicals to the smog. Studies have shown that wildfires contain the same deadly chemicals as structural wildfires, potentially releasing tons of gaseous and particulate pollutants. (Kim et al., 2018) Concerning the 2016 Fort McMurray

Wildfire, a study conducted by the Government of Alberta identified the potential for acute health effects from benzene exposure during the wildfire and acetaldehyde exposure after the wildfire. (Wentworth et al., 2018) Subsequently, another study was conducted by the University of Alberta to determine the effects on the respiratory health of firefighters attending the wildfire, concluding with the determination that exposures experienced during the wildfire are associated with non-resolving damage to health, notably the airways. (Cherry et al., 2021) In normal circumstances, firefighters use the air in SCBA tanks that might last up to an hour, providing enough time to work a single house fire but not enough for working multiple days in a row, fighting fires consuming thousands of homes and businesses. Forest crews sometimes wear particulate filter masks, which are not stocked in large numbers at municipal fire stations. Pallets of PPE arrived a few days after the fire had damaged the city. Working with filter masks is not ideal, as the masks make breathing harder, can get plugged, and may cause safety glasses and visors to fog. (Fredericks, 2019)

Short-term exposure to air pollutants may cause changes in the distal parts of the lungs, which need to be detected early. Short-term acute exposure to wildfire-related air pollutants is associated with lowered small airway function. The function of airways may be affected due to short-term acute wildfire-related smoke exposure due to a wide range of gaseous and particulate matter. (Moitra et al., 2021) Studies on general populations exposed to wildfires show long-term health effects. There is evidence of increased population-level mortality due to wildfire exposure and increased respiratory morbidity. Exposure to particulate matter and chemicals in wildfire smoke correlates with increased cancer risk. (Grant & Runkle, 2022) Surgical masks and respirators can provide limited protection for children during wildfire events, with expected decreases of roughly 20% and 80% for surgical masks and N95 respirators, respectively. (Holm et al., 2020) In a group of rhesus monkeys exposed in infancy to California wildfire smoke, lung function during adolescence decreased in the entire group, and inflammatory markers were also increased. This study points to the potential for lifelong health impacts from exposure to wildfire smoke early in life. (Black et al., 2017) The 2020 California Wildfire season was unique because it occurred during the COVID-19 pandemic, bringing further questions about how wildfire smoke affected human health and if the health effects of COVID-19 exasperated the damage caused by smoke inhalation and vice versa. A study conducted by the Environmental Systems Research Institute in Redlands, California, displayed strong evidence of positive associations between daily increases in particulate matter pollution and increased risks of COVID-19 cases and deaths. In certain counties, the percentage of COVID-19 cases and deaths attributable to the high levels of air pollution was substantial.

In many counties, the high levels of a particular matter that occurred during the 2020 wildfires substantially exacerbated the health burden of COVID-19, resulting in a concurrent event. (Zhou et al., 2021) If fitted and worn correctly, a federally certified N95 mask filters out 95 percent of particles larger than 0.3 microns, making N95 masks very efficient with keeping out the 2.5-micron particles found in wildfire smoke. Cloth masks cannot efficiently filter out the hazardous particles in wildfire smoke. They also often have gaps around the perimeter where polluted air may seep in, be inhaled, and enter the bloodstream via the lungs. Cloth masks are beneficial but not as good as an N95 mask when it comes to wildfire smoke, though any mask is better than no mask. (Ries, 2021)

While advocating for N95 masks may seem simple, there are many challenges to entire masking communities during an extended event, crisis, or emergency. The Fort McMurray wildfire began on May 1st, 2016, and forced the evacuation of the townsite by May 4th. In 2018, the infamous Camp Fire destroyed large sections of Paradise, California,

an event captured by the Netflix documentary "Fire in Paradise." Within one day, the fire grew by 5,000 acres in just three hours, expanding by an average of more than one American football field every 3 seconds. The most significant growth period occurred on a Thursday afternoon when it grew 10,000 acres in about 90 minutes, burning the equivalent of more than one football field every second. Due to the fire's unprecedented speed and growth, nine lives were lost. (Jones, 2018) Creating the logistics to import, store, deliver, and dispose of millions of masks is a monumental task that can not be feasibly achieved during acute emergencies. Given the variance in perception towards the risk of wildfire smoke, there may be little-to-no political will to provide free masks, especially to a general public that is burnt-out on pandemic messaging. Key interventions to protect cardiovascular health against wildfire smoke are similar to those that protect against airborne illnesses. These mitigation measures may include identifying and educating vulnerable patients, reducing outdoor activities, creating cleaner air environments using air filtration devices and personal respirators, and aggressively managing chronic diseases with long-term health care solutions. (Hadley et al., 2022) COVID-19 has impeded the ability of the government and the private sector to respond to wildfire risk. The scale of the 2020 wildfire season in many parts of the West, where drought in the 2019 to 2020 rainy season followed an accumulation of fuels during a relatively wet 2018 to 2019 season, has presented particularly acute challenges. Wildland firefighter training was delayed or sometimes canceled. Convict firefighter crews were unavailable due to early release from state prisons to avoid COVID outbreaks. Many fuel management treatments, such as prescribed burning or removing flammable materials like brush or debris, did not occur in winter and spring. Traditional approaches to wildfire evacuation have proved more challenging due to lowered capacity at evacuation centers resulting from social distancing requirements.
(M. Burke et al., 2021)

From California to Alberta and across the world, there are volumes of published literature and studies detailing the effects of wildfire smog on human health for firefighters and specific demographics of the general public, such as children. In the next section, we will discuss the public perception of wildfire smog and how that may affect public health efforts.

Public Perception of Wildfire Smog

The public perception of risk towards any hazard, otherwise known as all-risks, varies from time to time and place. While most would assume that protecting oneself from smoke is common sense, there are many complicating factors in individual perception of the risk of wildfire smoke, resulting in health effects from continuous exposure to smoke, smog, and other sources of pollution. Wildfires continue to transform landscapes as previously undeveloped areas are populated with residential and commercial structures, impacting ecosystems and creating new risky landscapes. Within this context, the science of wildfire risk mitigation has experienced renewed and enhanced support among stakeholders. Risk mitigation measures have yet to find a purchase in the public's acceptance or involvement in this new role of and for fire. Identifying and better understanding the values associated with wildfires will allow decision-makers to develop more effective communication programs and policies. (Gordon et al., 2020) Understanding how people perceive and respond to emerging science regarding the risk of wildfire smoke has received little attention. Interpreting and incorporating human perceptions of threats from wildfire smoke is critical, as decision-makers need such information to mitigate smoke-related risks. Information may include knowledge regarding the source of air quality information, perception of wildfire smoke as a hazard, and smoke-related health experiences. (Fowler et al., 2019) Humans' response to threats depends on their interpretation of the risk, shaped by their experience, personal feelings and values, cultural beliefs and societal dynamics. Understanding how people interpret risks and choose actions based on their interpretations is vital to any strategy for disaster reduction. Access to information and the capacity for self-protection are distributed unevenly within specific populations, making trust a critical moderator of the effectiveness of any policy for risk communication and public engagement. (Richard Eiser et al., 2012)

Predicting wildland fire smoke exposure is crucial to risk reduction, and this must occur in sync with communication about protective health resources. Public health and healthcare professionals rely on effective and factual communication strategies regarding risk, resulting in risk reduction strategies. For example, the COVID-19 pandemic was a critical concurrent public health risk during recent wildfire seasons. Unfortunately, the collective response to COVID-19 may jeopardize a proactive approach to wildfire smoke because of changing public perception, individual priorities, and limited resources to pay for further healthcare expenditures. Environmental health scientists are challenged to define at-risk populations, thus providing data and tools to support healthcare information and decisions. (Hagler et al., 2021) Many parts of North America have experienced extreme wildfire smoke in recent years. The number of studies on wildfire smoke and its acute health effects is small compared with the broader literature on ambient air pollution. The associations between exposure to particulate matter and all respiratory outcomes are clear and relatively large on smoky days. Evidence suggests that air pollution exposure leads to immunosuppression, inflammation, and increases in other respiratory symptoms. If COVID-19 continues to affect countries worldwide, smoke pollution may increase population susceptibility to the virus and cause more cases of severe disease. (Henderson, 2020) Quantifying the exact cost of wildfire smog mixing with respiratory diseases is nearly impossible. What may be done instead is the promotion of public health efforts to educate and create an understanding of the specific health risks of exposure to wildfire smoke.

Creating communication plans to increase the awareness of the public perception of wildfire smog may be difficult, but it is possible. Smoke risk communication is an essential part of community resilience to wildfires, helping to spread awareness of the health risks associated with smoke exposure and motivating the adoption of individual-level interventions to mitigate those health risks. Evidence suggests that delivery through trusted sources of information may improve the reception of risk communication messages. Specific communities or demographics may be hesitant to trust information from government agencies or major media outlets. Central themes centering on the importance of sources of information include the role of perceived credibility, the quality of information, and the level of trustworthiness between the public and private sectors. These themes should align with existing risk communication literature with detailed, qualitative descriptions from sources viewed as trustworthy. (Wood et al., 2022) Effective communication about the health effects of wildfire smoke is vital to protect the public, especially those most vulnerable to exposure, including people with chronic respiratory conditions, children, and seniors. To be successful, materials regarding the mitigation of wildfire smog should present specific behavioral recommendations. Suggestions may include avoiding exposure to air pollution by wildfire smoke by staying indoors, reducing activity levels, and using air purifiers or high-quality masks. However, materials are often needed to acknowledge any uncertainty around these recommendations. Creators of these materials may want to incorporate more relevant illustrations to support the main message and consider how information about the risks and benefits of the recommended behaviors can most clearly be presented. (Cook, 2020) Accurate risk perception and understanding of wildfire smoke are necessary for adherence to public health initiatives. There is importance towards the advocacy and implementation of education efforts to increase health perceptions and improve outcomes despite exposure to extended periods of wildfire smog. Further research is needed to minimize misperceptions and personal risk perceptions toward wildfire smoke. The goal should be to shape risk perception and effectively communicate facts regarding health effects accurately.

The successful implementation of health measures depends on public acceptance of these measures. Psychological variables, such as trust and worldviews, strongly influence people's risk perceptions and acceptance of mitigation efforts. Since the approval of implemented measures depends on whether they align with people's values and worldviews, the latter two variables are relevant considerations for successful risk management strategies. The COVID-19 pandemic has shown that as soon as the measures attain success or the public is tired of the implemented restrictions, public acceptance declines, and it seems complicated to prolong any efforts toward any risk. The importance of worldviews and trust for public acceptance of health measures may necessitate political discussions about the implemented measures. (Siegrist & Bearth, 2021) Lessons learned from COVID-19 communication strategies may influence practical mitigation efforts toward educating the public and stakeholders on the health effects of wildfire smoke.

Not only do medical institutions (hospitals, medical foundations, health authorities, etc.) play the role of trustworthy agents, but also healthcare personnel emerge as legitimate agents to lead communicative efforts. Using public health representatives to spread messaging may be the option for increasing public perception of risk. Citizens who perceive their public leaders as trustworthy are more likely to comply with government demands relevant to a particular crisis because an effective response depends on collective compliance with public health guidelines. Abundant instances of bounded solidarity in terms of altruism, gratitude and reciprocity may increase during crises, be it pandemics or wildfires. Any public communication effort should prioritize finding synergies between this untapped social potential and governmental policies. (Yter et al., 2021) In parts of North America, it is possible that mask mandates towards any form of risk will never be enacted again. The political capital to force individuals to mask may never exist again for certain politicians. Public communication efforts at caring for the well-being of healthcare systems may need to consider the political climate and capability of weary communities to consume actionable information regarding wildfire smoke. Public health messages must strive for consistency and clarity, especially after the often confusing and changing information regarding the COVID-19 pandemic.

Consistency in public health recommendations caused confusion, frustration and mistrust, laid the ground for mask skepticism and resulted in friction between those wearing masks and those choosing to go without masks. To combat the adverse effects of change, shifts in public health messages should be clearly explained and accompanied by the best available evidence to help the public understand why the changes were necessary. As mask use extends, public health messaging for mask literacy needs to continue, with prioritization on the honest and transparent flow of information with the public on the benefits and precautions in using specific preventative measures, mainly concerning personal safety and protection from wildfire smoke. Incorporating these areas of knowledge into future health messaging could help individuals better understand the recommended actions, guiding them to make more informed choices in preventative health practices. (Zhang et al., 2021) Messaging needs to be informative and positive and should include references or citations. Understanding that the public may be weary of messaging should not be a hindrance but an opportunity to create new and unique material.

In conclusion, there is a continued need to educate members of the public, including first responders and healthcare authorities, on the health effects of wildfire smoke. There may be increased difficulty in these efforts due to public hesitation for masking, weariness toward government interventions, and general malaise towards any health interventions due to the COVID-19 pandemic. As scientific evidence of harm to human health from wildfire smoke and smog becomes more apparent, stakeholders' responsibility to communicate the risks to human health increases. Creating timely, relevant, factual, and easy-to-consume communication materials is challenging but entirely achievable in our modern media environment.

References

Black, C., Gerriets, J. E., Fontaine, J. H., Harper, R. W., Kenyon, N. J., Tablin, F., Schelegle, E. S., & Miller, L. A. (2017). Early Life Wildfire Smoke Exposure Is Associated with Immune Dysregulation and Lung Function Decrements in Adolescence. *American Journal of Respiratory Cell and Molecular Biology*, *56*(5), 657–666. https://doi.org/10.1165/rcmb.2016-0380oc

Burke, M., Driscoll, A., Heft-Neal, S., Xue, J., Burney, J., & Wara, M. (2021). The changing risk and burden of wildfire in the United States. *Proceedings of the National Academy of Sciences*, *118*(2). https://doi.org/10.1073/pnas.2011048118

Burke, R. (2016, December 31). *Hydrogen Cyanide: The Real Killer Among Fire Gases*. https://www.firehouse.com/. Retrieved December 13, 2022, from https://www.firehouse.com/rescue/article/10502165/hydrogen-cyanide-the-real-killer-among-fire-gases

CalFire. (2022). *2020 Fire Season*. https://www.fire.ca.gov/. Retrieved December 13, 2022, from https://www.fire.ca.gov/incidents/2020/

Cherry, N., Barrie, J. R., Beach, J., Galarneau, J. M., Mhonde, T., & Wong, E. (2021). Respiratory Outcomes of Firefighter Exposures in the Fort McMurray Fire. *Journal of Occupational &Amp;*

Environmental Medicine, 63(9), 779–786. https://doi.org/10.1097/
jom.0000000000002286

Cook, L. A. (2020). Communicating the health risks of wildfire
smoke exposure: Health literacy considerations of public
health campaigns. *American Thoracic Society.* https://doi.
org/10.1101/2020.09.14.20194662

Fowler, M., Modaresi Rad, A., Utych, S., Adams, A., Alamian, S.,
Pierce, J., Dennison, P., Abatzoglou, J. T., AghaKouchak, A., Montrose,
L., & Sadegh, M. (2019). A dataset on human perception of and response
to wildfire smoke. *Scientific Data, 6*(1). https://doi.org/10.1038/s41597-
019-0251-y

Fredericks, C. (2019, May 21). *Fort Mac firefighters face potential
health issues.* Canadian Firefighter Magazine. https://www.cdnfirefighter.
com/fort-mac-firefighters-face-potential-health-problems-40538/

Gordon, J., S. Willcox, A., Luloff, A., C. Finley, J., & G. Hodges,
D. (2020). Public Perceptions of Values Associated with Wildfire
Protection at the Wildland-Urban Interface: A Synthesis of National
Findings. *Landscape Reclamation - Rising From What's Left.* https://doi.
org/10.5772/intechopen.82171

Grant, E., & Runkle, J. D. (2022). Long-term health effects of wildfire
exposure: A scoping review. *The Journal of Climate Change and Health,
6,* 100110. https://doi.org/10.1016/j.joclim.2021.100110

Hadley, M. B., Henderson, S. B., Brauer, M., & Vedanthan, R. (2022).
Protecting Cardiovascular Health From Wildfire Smoke. *Circulation,
146*(10), 788–801. https://doi.org/10.1161/circulationaha.121.058058

Hagler, G. S. W., Henderson, S. B., McCaffrey, S., Johnston, F. H., Stone, S., Rappold, A., & Cascio, W. E. (2021). Editorial: Understanding and Communicating Wildland Fire Smoke Risk. *Frontiers in Public Health, 9*. https://doi.org/10.3389/fpubh.2021.721823

Henderson, S. B. (2020). The COVID-19 Pandemic and Wildfire Smoke: Potentially Concomitant Disasters. *American Journal of Public Health, 110*(8), 1140–1142. https://doi.org/10.2105/ajph.2020.305744

Holm, S. M., Miller, M. D., & Balmes, J. R. (2020). Health effects of wildfire smoke in children and public health tools: a narrative review. *Journal of Exposure Science &Amp; Environmental Epidemiology, 31*(1), 1–20. https://doi.org/10.1038/s41370-020-00267-4

Holstege, C., Forrester, J., Borek, H., & Lawerence, D. (2015, March 13). *A Case of Cyanide Poisoning and the Use of Arterial Blood Gas Analysis to Direct Therapy*. https://www.tandfonline.com/. Retrieved December 13, 2022, from https://www.tandfonline.com/doi/abs/10.3810/hp.2010.11.342?journalCode=ihop20

Jones, J. (2018, November 10). One of the California wildfires grew so fast it burned the equivalent of a football field every second. *CNN*. https://edition.cnn.com/2018/11/09/us/california-wildfires-superlatives-wcx/index.html

Julia Parrish. (2018, May 4). *Timeline: 'The Beast' hits Fort McMurray, and the recovery*. Edmonton. https://edmonton.ctvnews.ca/timeline-the-beast-hits-fort-mcmurray-and-the-recovery-1.3914168

Kim, Y. H., Warren, S. H., Krantz, Q. T., King, C., Jaskot, R., Preston, W. T., George, B. J., Hays, M. D., Landis, M. S., Higuchi, M., DeMarini, D. M., & Gilmour, M. I. (2018). Mutagenicity and Lung Toxicity of Smoldering vs. Flaming Emissions from Various Biomass Fuels: Implications for Health Effects from Wildland Fires. *Environmental Health Perspectives, 126*(1), 017011. https://doi.org/10.1289/ehp2200

McDermott, V. (2022, December 10). *NDP propose full presumptive WCB coverage for firefighters that fought 2016 wildfire.* Fortmcmurraytoday. https://www.fortmcmurraytoday.com/news/ndp-propose-full-presumptive-wcb-coverage-for-firefighters-that-fought-2016-wildfire

Moitra, S., Tabrizi, A. F., Fathy, D., Kamravaei, S., Miandashti, N., Henderson, L., Khadour, F., Naseem, M. T., Murgia, N., Melenka, L., & Lacy, P. (2021). Short-Term Acute Exposure to Wildfire Smoke and Lung Function among Royal Canadian Mounted Police (RCMP) Officers. *International Journal of Environmental Research and Public Health, 18*(22), 11787. https://doi.org/10.3390/ijerph182211787

Richard Eiser, J., Bostrom, A., Burton, I., Johnston, D. M., McClure, J., Paton, D., van der Pligt, J., & White, M. P. (2012). Risk interpretation and action: A conceptual framework for responses to natural hazards. *International Journal of Disaster Risk Reduction, 1*, 5–16. https://doi.org/10.1016/j.ijdrr.2012.05.002

Ries, J. (2021, August 3). *Will Your COVID-19 Mask Protect You from Wildfire Smoke?* Healthline. https://www.healthline.com/health-news/will-your-covid-19-mask-protect-you-from-wildfire-smoke

Siegrist, M., & Bearth, A. (2021). Worldviews, trust, and risk perceptions shape public acceptance of COVID-19 public health measures. *Proceedings of the National Academy of Sciences, 118*(24). https://doi.org/10.1073/pnas.2100411118

Wentworth, G. R., Aklilu, Y. A., Landis, M. S., & Hsu, Y. M. (2018). Impacts of a large boreal wildfire on ground level atmospheric concentrations of PAHs, VOCs and ozone. *Atmospheric Environment*, *178*, 19–30. https://doi.org/10.1016/j.atmosenv.2018.01.013

Wood, L. M., D'Evelyn, S. M., Errett, N. A., Bostrom, A., Desautel, C., Alvarado, E., Ray, K., & Spector, J. T. (2022). "When people see me, they know me; they trust what I say": characterizing the role of trusted sources for smoke risk communication in the Okanogan River Airshed Emphasis Area. *BMC Public Health*, *22*(1). https://doi.org/10.1186/s12889-022-14816-z

Xu, R., Yu, P., Abramson, M. J., Johnston, F. H., Samet, J. M., Bell, M. L., Haines, A., Ebi, K. L., Li, S., & Guo, Y. (2020). Wildfires, Global Climate Change, and Human Health. *New England Journal of Medicine*, *383*(22), 2173–2181. https://doi.org/10.1056/nejmsr2028985

Yter, M., Murillo, D., & Georgiou, A. (2021). Bounded Solidarity as an Asset for Public Health Care Intervention. *Qualitative Health Research*, *32*(3), 440–452. https://doi.org/10.1177/10497323211057081

Zhang, Y. S. D., Young Leslie, H., Sharafaddin-zadeh, Y., Noels, K., & Lou, N. M. (2021). Public Health Messages About Face Masks Early in the COVID-19 Pandemic: Perceptions of and Impacts on Canadians. *Journal of Community Health*, *46*(5), 903–912. https://doi.org/10.1007/s10900-021-00971-8

Zhou, X., Josey, K., Kamareddine, L., Caine, M. C., Liu, T., Mickley, L. J., Cooper, M., & Dominici, F. (2021). Excess of COVID-19 cases and deaths due to fine particulate matter exposure during the 2020 wildfires in the United States. *Science Advances*, *7*(33). https://doi.org/10.1126/sciadv.abi8789

The Health Effects of Wildfire Smoke

Chapter 8: Paying the Price

JJ Doleweerd

Introduction

Wildfires are uncontrollable in their nature. The effects can range from having a minimal impact to an absolutely devastating one. These fires that start small grow into massive fires that cause destruction, pain and suffering to anything and anyone in their path. They inflict a combined global toll on our planet and the species that live upon it. It is the aim of this chapter to discuss the toll these fires inflict on humans as well as the financial toll that it places on society. Wildfires are a part of nature and have been playing a significant role in maintaining balance in ecosystems and in eradicating pests (Pausas and Keeley, 2019). As the climate continues to change as a result of global warming, the frequency and severity of wildfires are only going to increase, so it is vital to establish the current tolls these fires take on society so that the world can be properly motivated to invest in the prevention and containment of these fires. Not only that but these "fire emissions are an important contributor to global mortality" (Johnston et al., 2012) and we should be taking steps to prevent the adverse effects on health in a global sense as a result of climate change but also take steps at the acute, local level as well as keeping global economies stable.

There are two main avenues in which wildfires inflict damage and harm on a region and the world that this chapter will explore. These are the human toll and the financial toll of wildfires. The human toll will discuss the variety of ways these fires impact humans individually and as a population. It will also discuss how the severity and effect of wildfires vary between developed and developing nations. The economic toll will explore the adverse effects these wildfires have on economies at both the local and global levels, as well as the industries that are impacted the most and are directly affected by wildfires.

Human Toll

Wildfires take a toll on humans in many ways. There are injuries and loss of life that they result in. In addition, there is the effect that these fires have on climate change, which creates risks to the health and well-being of humans across the globe. There is also the risk they pose to the surrounding region from the reduction in air quality from the smoke produced and the risk the air pollution and smoke inhalation present. Aside from the health risks, there are infrastructure losses, the displacement of people from their houses, unemployment from the businesses struggling because of these fires, and a reduction in quality of life from the combined devastation to the region.

One recent case that can be examined to get a better and more holistic understanding of how wildfires take a toll on human populations is the Australian Bushfire of 2019-2020. Another case that will be examined is the 2008 fires on the savannah in Botswana. This case will contrast the abilities of developed nations to adapt and recover from a devastating natural event versus that of a developing nation.

Australian Bushfire of 2019–2020

Climate change has increased the frequency of drought which increases the risk of wildfires by making conditions ideal for a fire to spread. This has led to an increased number of wildfires in recent years, spanning larger areas and having greater impacts. One of the major tolls these wildfires take is on human health. The air pollution caused by the smoke produced has negative impacts on respiratory and cardiac function (Rodney et al., 2021). As seen in the catastrophic Australian bushfires of 2019-2020, one survey found that 97 percent of respondents attributed at least one symptom to the bushfire smoke produced (Rodney et al., 2021). In addition, it is estimated that in these catastrophic bushfires, over 2400 homes were destroyed by these fires (Rodney et al., 2021). As the frequency of wildfires increases, similar events will continue to occur if preventative measures are not taken to minimize the fires and their damages.

It is well understood that the burning of fuels releases greenhouse gases such as carbon dioxide into the atmosphere, and the smog produced pollutes the air we breathe. Wildfires can have devastating effects on a region not only because of the destructive and consuming nature of fire but also because the emissions from fire play a significant role through their contributions to global mortality (Johnston et al., 2012). It has been observed that "fires across the U.S. in 2019 caused about 3,700 civilian deaths and another recorded 16,600 injuries" (Reiff, 2022). In addition, there are other factors of health that are to be considered. Many people are traumatized by the experience of surviving a wildfire, from injuries they have sustained as a result, or for other reasons because all they know might have just been burned down (Reiff, 2022).

2008 Botswana Wildfires

Another recent example of the devastation that wildfires have caused on a society and its economy takes place on the savannah of Botswana in 2008. The fire caused by abnormal weather devastated the local economy dependent on thatch collection and tourism (The New Humanitarian, 2015). Already, the capacity of this developing country to contain the fire or recover from the effects of a fire is lower than that of a developed, wealthy country. When the economy then crashes, this has unimaginable economic consequences which lead to a significant reduction in the average quality of life of a citizen there. Many people would have died from the fires themselves but also as a result of the financial toll these fires would have on individuals and their families. Not only do wildfires have such a great economic impact on developing countries, but it was also found that other developing countries, specifically in sub-Saharan Africa, with Botswana as an example, and Southeast Asia have the highest mortality rates (Johnston et al., 2012).

Often, wildfires are started as controlled fires, like a campfire, but they are poorly controlled and contained such that they spread and grow into uncontrollable fires, each would then be known as a wildfire. To work towards the prevention of future wildfires occurring, the public must be educated in fire safety so they can learn how to better keep recreational fires controlled. However, developing countries like Botswana face further difficulties than many other developed countries. Currently, Botswana's fire prevention and mitigation strategies are centered around government action (Dube, 2013). This seems ineffective, and it is proposed that they shift these strategies to a more community-inclusive approach for greater efficacy (Dube, 2013). A community-inclusive approach would make more sense because actions could be taken more quickly and in a more precise manner than the broad strokes government policy often results in.

Health Impacts

While the immediate area around a wildfire faces the risk of injury and death from the fire and destruction itself, wildfires can affect the health of people in a much larger range in other ways. The smoke contributes to climate change which has drastic effects on health, but the smoke and air pollution created and emitted by the fires can be extremely harmful to the lungs, for people even thousands of miles away (American Lung Association, 2016). It is apparent that smoke inhalation will increase the risk of cancer, however, studies have also shown that long-term exposure to wildfires would also increase the risk of developing a brain tumor by 10 percent (Deibert, 2022). In addition, long-term exposure to wildfires also increases the risk of lung cancer by 5 percent (Deibert, 2022). This major health complication makes the increasing rate of wildfires all the more concerning.

Not only do wildfires have consequences on the physical health of humans, but mental health is also affected negatively. Current research data suggests that the victims of wildfires are at an increased risk of suffering from post-traumatic stress disorder or depression (Caamano-Isorna et al., 2011). Studies have found that 33 percent of victims are "probable major depression three months after the traumatic event" (Caamano-Isorna et al., 2011). As the rate of wildfires continues to increase, so will the number of people struggling with their mental and physical health.

Financial Impacts

Wildfires are sources of immense destruction. This destruction has many rippling effects on the economy, including land loss and property damage to the costs of the negative effects they have on human health. Some industries are also impacted much more significantly than other industries because of the immediate negative effects that wildfires cause.

Land Loss

There have been significant losses and damage done not only to human health but financially, to our ecosystems and to the environment. In the United States alone, "since 2000, an annual average of seven million acres has burned, reflecting the rapid proliferation of wildfires across the country. As of August 2022, 48,211 wildfires have burned nearly 6.2

million acres, on track to surpass last year's devastation." (Reiff, 2022). This damage will be magnitudes of order greater if the entire planet is considered. Many developing countries are much more prone to wildfires and have significantly fewer resources to manage and contain these fires than the United States and other developed countries as well.

Industries Impacted

Some industries have less resilience to wildfires than other industries because they have a dependency on various resources, property or land that can be devastated and destroyed in a wildfire. Another impact that wildfires can have on businesses is that the safety hazards drive people away from businesses that rely on customers being there in person. The main industries that are affected and will be discussed in this chapter are tourism, forestry, agriculture, the housing and real estate industry and the insurance industry.

As wildfires inflict destruction on a region they not only destroy property and businesses but also impact tourism in the region negatively. In the case of the tourism industry, this industry is dependent on certain factors such as the safety of the region. Tourists will be less likely to visit a region that is suffering from a wildfire because doing so would unnecessarily endanger their health (Reiff, 2022). Another reason wildfires pose a great risk to regions that have economies which rely on tourism is that these massive fires prevent people from coming to the region and often destroy the very nature which is what the tourists are coming to see (Reiff, 2022). Essentially, "tourism and recreation are decimated" (Reiff, 2022). For countries like Botswana, with economies dependent on tourism, wildfires are catastrophic events because they have the potential to completely ruin an economy.

Other industries such as agriculture and forestry also face significant impacts from wildfires. They risk losing significant value in property loss through wildfires because of the land that is burned and in forestry's case specifically, the trees in forests that are burned down and can no longer be utilized (Stephenson et al., 2012). It can make it very hard for businesses in these industries to recover from the financial losses that they have experienced. Similarly to forestry, the agriculture industry also has immovable assets that are critical to their income. Crops would be decimated if they find themselves in the path of a wildfire. These are irrecoverable assets that can have significant adverse effects on the farm's ability to support itself financially that year. Given the nature of farming, crops destroyed are completely lost for the entire growing season, in addition to the negative effects, the burning can have on the ability of the farmland to grow crops in future years (Benchmark Labs, 2022). This is very apparent when a wildfire burns through even just part of an orchard. The ability to produce crops and that rate is severely impacted for the following years until new trees are grown and able to fill the void (Benchmark Labs, 2022).

The housing and real estate industry is also greatly impacted by wildfires. As the rate of wildfires continues to increase, there is new legislation being passed that adds increasing restrictions and requirements to new buildings to reduce the likelihood of being destroyed by the fires and further fueling them (Rossi, 2014). This will increase the cost to construct new houses and buildings further driving the cost of real estate upwards and increasing the cost of living and the cost of renting property. The insurance industry exists to support people in the event of a disaster striking. This means that the insurance industry will bear a significant portion of the costs that are incurred from wildfires. As a result of this, premiums on insurance will rise so that they are able to cover the increase in insurance payouts because of the wide scale of damage that wildfires cause. Homes in fire-prone areas will have higher insurance premiums because the risk of damage is so much higher (Hazra & Gallagher, 2022). This directly impacts businesses and homeowners by increasing their costs and can also contribute to the rising prices of properties in housing markets. It is clear that insurance has a wide reach into other industries because of its importance and prevalence in society and as a result negative impacts on the insurance industry will have far reaching consequences on other industries.

Property Damage

The financial impacts that wildfires inflict through property damage are also significant. In the United States in 2020, fires "caused $21.9 billion in property damage" (Reiff, 2022). There are many fires that take place each year that contribute to this total. However, each of the larger individual wildfires alone can cause more than 1 billion dollars (Reiff, 2022). Again, this is magnitudes of order greater when scaled to consider the property damage inflicted on the rest of the planet because of wildfires. The Camp Fire took place in November 2018 in California and is the most destructive wildfire on record there to date, with the Tubbs fire taking place in October 2017 as the second most destructive wildfire there. It was estimated that "the value of property loss at $11 billion to $13 billion from the campfire and $5 billion to $7 billion from the Tubbs Fire." (Hepp, 2021). To better understand the impact these fires had, the Camp Fire damaged or destroyed 18,800 structures (Hepp, 2021). These structures would include businesses and houses resulting in significant displacement of the people in the town and businesses shutting down, unable to recover from the damages they incurred.

Prevention

Clearly, further measures should be taken to prevent wildfires, thus reducing the toll these fires take on humans, their health, businesses and the economy. They cause such large disruptions that there should be increased efforts to maintain stability and reduce the frequency of these large-scale disasters.

It is obvious wildfires have a significant impact at a local and regional scale because of the nature of their destruction, however, the issue of the increased rate of wildfires must be addressed more impactfully at a local level. This is a global problem that requires a global effort toward a solution. Of course, reducing the contributions everything and everyone makes to worsening climate change and greenhouse gasses is a great start because of the role that climate change plays in setting the stage for a wildfire to occur. The risk of drought-induced fires is increasing as a direct result of climate change and global warming (Balling et al., 1992). This is because as the land becomes increasingly arid, with increasing temperatures, it sets the stage for large-scale wildfires to occur much more frequently (Balling et al., 1992). This demonstrates the increased importance of making more efforts that minimize climate change, greenhouse gas emissions and global warming in regard to wildfires.

Another viable method to reduce the risk and impacts of wildfires is to motivate and educate the public on wildfire mitigation. This is crucial because 50 percent of Canadian wildfires are caused as a direct result of human action (Tymstra et al., 2020). It is clear public education will play a significant role in reducing the rate at which wildfires occur. However, public education is a much harder task in developing countries where there is less existing infrastructure in place for public education programs of any type. This is why this last method of wildfire prevention will play a significant role in wildfire mitigation. Fuel management is an important method to be considered when aiming to prevent wildfires from occurring. Fuel management controls the availability of fuel that could be consumed in an area if a fire were to take place. In the United States, this means "the strategic removal of grasses, shrubs, and trees to restore and maintain ecosystems and limit the negative impacts of wildfires" (U.S. Department of the Interior, 2022). This can help control fires because these wildfires cannot grow so large and become as devastating with the necessary fuel. It is also more applicable to all countries because any country can begin to make efforts with urban planning and include fuel management as a consideration when creating regulations and new legislation. This will keep more people in their houses and businesses afloat if they are not suffering the same devastation from wildfires because it would be significantly harder for any fire to spread.

Conclusion

Wildfires take a massive toll on our world, humans, and our economies, devastating industries and businesses, and causing a large number of impacts on physical and mental health that have also resulted in the loss of life. There is hope that the frequency of wildfires can be reduced through the few methods of many that were discussed in this chapter. Additionally, fuel management would also help in reducing the impact each wildfire would have if it is implemented. The massive cost these fires take on the planet needs to be well understood to gain motivation and understanding from the public to take action toward wildfire mitigation. It makes sense from a financial and societal perspective, as it is advantageous to invest in preventing the damage from being inflicted in the first place rather than be forced to spend money to replace, repair and recover societally from the damage. This is why wildfires need to be given more attention because of the toll they are increasingly taking on human health and economies.

References

Balling, R. C., Meyer, G. A., & Wells, S. G. (1991, December 5). *Climate change in Yellowstone National Park: Is the drought-related risk of wildfires increasing?* SpringerLink. Retrieved December 30, 2022, from https://link.springer.com/article/10.1007/BF00143342#citeas

Benchmarklabs. (2022, April 6). *How do wildfires affect agriculture.* Benchmark Labs. Retrieved December 30, 2022, from https://www. benchmarklabs.com/blog/how-do-wildfires-affect-agriculture/

Caamano-Isorna, F., Figueiras, A., Sastre, I., Montes-Martinez, A., Taracido, M., & Pineiro-Lamas, M. (2011, May 21). *Respiratory and mental health effects of wildfires: An Ecological Study in Galician municipalities (North-West Spain) - environmental health.* SpringerLink. Retrieved December 30, 2022, from https://link.springer.com/ article/10.1186/1476-069X-10-48

Deibert, E. (2022, June 9). *The risk of cancer is spreading like wildfire.* Research2Reality. Retrieved December 30, 2022, from https:// research2reality.com/health-medicine/cancer/wildfire-cancer-risks-climate-change-danger/

Developing countries hardest hit by wildfires. The New Humanitarian. (2015, November 3). Retrieved December 30, 2022, from https://www. thenewhumanitarian.org/report/93072/global-developing-countries-hardest-hit-wildfires

Dube, O. P. (2013, August 21). *Challenges of wildland fire management in Botswana: Towards a community inclusive fire management approach.* Weather and Climate Extremes. Retrieved December 30, 2022, from https://www.sciencedirect.com/science/article/pii/ S2212094713000091#bib3

Fuels management. U.S. Department of the Interior. (2022, April 20). Retrieved December 30, 2022, from https://www.doi.gov/wildlandfire/ fuels

Hazra, D., & Gallagher, P. (2022, January 13). *Role of insurance in wildfire risk mitigation.* Economic Modelling. Retrieved December 30, 2022, from https://www.sciencedirect.com/science/article/abs/pii/S0264999322000141

Hepp, S. (2022, August 17). *The impact of wildfires on Rent & Home Prices.* CoreLogic®. Retrieved December 30, 2022, from https://www.corelogic.com/intelligence/the-impact-of-wildfires-on-rent-home-prices/

How wildfires affect our health. American Lung Association. (2016, January 1). Retrieved December 30, 2022, from https://www.lung.org/blog/how-wildfires-affect-health

Johnston, F. H., Henderson, S. B., Chen, Y., Randerson, J. T., Marlier, M., DeFries, R. S., Kinney, P., Bowman, D. M. J. S., & Brauer, M. (2012, May 1). *Estimated global mortality attributable to smoke from landscape fires.* National Institute of Environmental Health Sciences. Retrieved December 30, 2022, from https://ehp.niehs.nih.gov/doi/full/10.1289/ehp.1104422

Pausas, J. G., & Keeley, J. E. (2019, May 6). *Wildfires as an ecosystem service.* Retrieved December 31, 2022, from https://esajournals.onlinelibrary.wiley.com/doi/10.1002/fee.2044

Reiff, N. (2022, November 22). *How fire season affects the economy.* Investopedia. Retrieved December 30, 2022, from https://www.investopedia.com/how-fire-season-affects-the-economy-5194059

Rodney, R. M., Swaminathan, A., Calear, A. L., Christensen, B. K., Lal, A., Lane, J., Leviston, Z., Reynolds, J., Trevenar, S., Vardoulakis, S.,

& Walker, I. (2021, October 14). *Physical and Mental Health Effects of Bushfire and Smoke in the Australian Capital Territory 2019–20.* Frontiers. Retrieved December 30, 2022, from https://www.frontiersin. org/articles/10.3389/fpubh.2021.682402/full

Rossi, A. J. (2014). *Wildfire Risk and the Residential Housing Market: A Spatial Hedonic Analysis.* University of Pennsylvania. Retrieved December 30, 2022, from https://repository.upenn.edu/cgi/viewcontent. cgi?article=1212&context=curej

Stephenson, C., Handmer, J., & Betts, R. (2012, October 12). *Estimating the economic, social and environmental impacts of wildfires in Australia.* Taylor & Francis. Retrieved December 30, 2022, from https://www. tandfonline.com/doi/full/10.1080/17477891.2012.703490

Tymstra, C., Stocks, B. J., Cai, X., & Flannigan, M. D. (2019, October 31). *Wildfire Management in Canada: Review, Challenges and Opportunities.* Progress in Disaster Science. Retrieved December 30, 2022, from https://www.sciencedirect.com/science/article/pii/ S2590061719300456

Chapter 9: Investing in Climate Change

Christina MacDonald

Understanding Climate Change

To truly understand the importance of globally investing in combating climate change, one must understand a few things about climate change. This section will delve into some important terms and the global implications of climate change. It will also provide a brief overview of why climate change is a global issue and a brief look into the history of understanding climate change and climate change activism to lay the foundations for understanding the importance of investing in climate change. Unless otherwise indicated, all dollar amounts will be in United States Dollars.

Climate Inaction, Climate Action and Climate Adaptation

This chapter will use terms such as climate inaction, climate action, and climate adaptation. Climate inaction refers to not taking action in order to combat climate change whereas climate action refers to the opposite—taking necessary steps and implementing strategies to fight climate change. Climate adaptation can be linked with climate action as it is the implementation of strategies to help populations live by adapting to the environmental changes caused by climate change

(United Nations [UN], 2021a). The UN (2021a) provides examples of small changes, such as planting trees on one's property to keep the home cooler or acquiring insurance relevant to the natural disasters the area in which one resides is likely to experience. These changes do not have to directly slow down global warming but can be a response to it that helps individuals protect themselves and their livelihoods. On a larger level, the UN (2021a) talks about how governments play a role in protecting the economy and society. They provide examples such as governments ensuring there are safety measures in place like bridges and roads that remain unharmed in higher temperatures or systems in cities that help to mitigate flooding damage.

Climate Change on a Global Level

Climate change is a global issue because it impacts not only the resources that are available but the peace and well-being of populations and nations around the world. Although it is important to be conscious of one's impact on the environment, climate change is a global issue that must be addressed by organizations of power, such as world governments. In their article entitled, *The climate crisis – A race we can win,* the UN (2019) writes about how climate change not only causes environmental damage but causes major problems regarding world peace and resource security. With necessary resources like food, water, and land becoming less abundant, there is more global competition over them (UN, 2019). This results in more people needing to move to find better living circumstances, referred to as *mass displacement.* Returning to the concept of climate adaptation, sometimes relocating is the only way to adapt to the changes that are being faced by a community (UN, 2021a).

The UN (2019) reports that "90 percent of disasters are now classed as weather- and climate-related, costing the world economy" 520 billion dollars every year. The article also states that 26 million people are impacted, falling into poverty due to these issues. As will be explored later on in the chapter, investing in changes early can help save money and lives later on (UN, 2021).

A Brief Look Into the History of Fighting Climate Change

There was a time when the warming of the Earth was considered a favorable outcome; it was thought that if the Earth became warmer, people would enjoy the climate more, especially if they lived in areas of the world that were much colder (History.com Editors, 2022). This frame of thinking dates back to the late 1800s, however, it was not until the late 1900s—1989 to be exact—that there was a monumental shift in climate change strategies as more action started to be taken on a global level to address its impact as well as strategies to combat it (Jackson, 2007; History.com Editors, 2022). This was not the first time topics such as pollution, use of the Earth's resources or other environmental issues were relevant, but the 1980s was a significant time because there was a noticeable change in temperatures globally (History.com Editors, 2022).

Discussions and debates around climate change have been extremely prominent in recent years. Names such as David Suzuki, Al Gore, and Great Thunberg are associated with global warming and climate change activism. More specifically, David Suzuki—a Canadian environmental activist who has several other accolades, including radio show host, television show host, and author—has been involved in climate activism for decades (Eldridge, 2022; David Suzuki Foundation n.d.). Notably, his foundation's website writes about him educating the public about the dangers of global warming in 1989 and starting the David Suzuki Foundation in 1990 (David Suzuki Foundation n.d.). The David Suzuki Foundation (n.d.) aims to spread awareness, inspire leadership, and encourage positive change by finding solutions to pressing environmental issues. Moreover, Al Gore, a former vice president of the United States of America and winner of a Nobel Prize in 2007 for his climate change work, founded the Climate Reality Project in 2005 which aims to educate individuals about the negative outcomes of global warming and climate change (Al Gore, n.d.; NobelPrize.org, n.d.). The project advertises that it currently has thousands of representatives, has trained thousands of individuals to be better leaders in the fight against climate change and actively highlights how climate change is impacting different countries around the world (The Climate Reality Project, n.d.). Lastly, Greta Thunberg, named *TIME*'s Person of the Year in 2019, gained worldwide popularity for her climate activism. From engaging in a climate protest by sitting outside of Parliament in Sweden to being a leader in the largest demonstration regarding climate in all of history, her messages about the need for climate change have reached millions (Alter et al., 2019). These are only a few of the widely publicized climate activists and projects—stemming from Canada, the United States, and Sweden—so it does not even fully represent the global efforts being taken to combat climate change. Although they have impacted climate activism around the world, there have been several other important activists and organizations that are devoted to inspiring positive change regarding environmental issues.

Even though this section only provided a brief look into historical and current discussions and actions surrounding climate change, it is clear that the world's understanding of climate change, the impact of climate change, and the strategies needed to combat it are incredibly nuanced.

Investing in Climate Change

Now that some background on the climate crisis and efforts to address it have been discussed, it is time to delve deeper into the financial aspect of combating climate change. To try to quantify the financial profits countries around the world would have combined is a very difficult task. There has been a lot of research and discussion around the costs and benefits of investing in climate change globally, but when delving deeper into the literature, some of the conversations seem to be more scientifically sound than others. Even in searching on the internet, writing something along the lines of "how much would it take to solve climate change", there are several different numbers that arise, making it difficult to truly understand the costs and large investments that would need to be involved. It is also made more complex when considering how not only one source contributes to climate change. This section will delve into some of the conversations around different cost estimates, but one must understand that the numbers are all estimates and projections and should not be taken as absolutes. Predictions for the costs and benefits of globally fighting climate change must be analyzed critically. It is extraordinarily difficult to conceptualize these incredibly large sums of money, so this section will look at various literature and articles that discuss the cost of climate action and climate inaction, but it will also try to put some of the numbers into perspective.

The Cost of Climate Inaction compared to Climate Action

As discussed earlier, there are several areas that climate change impacts, including agriculture, water supplies, natural disasters, displacement of very large groups of people and more. Another demand that will become even more costly due to climate change is electricity. McFarland et al. (2015) looked into the impact climate inaction would have on the supply and demand of electricity in the United States and found that there is a projected demand for electricity in connection with temperatures rising in the United States. Regarding climate inaction, the need for increased electricity would bring with it increased costs which they compared to being similar to the costs needed to reduce carbon emissions produced by the power sector by half in the same timeframe (McFarland et al., 2015). This highlights that humans have exacerbated the impacts of global warming, which are then creating more demands for resources (UN, 2019). Unfortunately, there is also a greater strain on resources the longer there are only minimal efforts to tackle climate change. As expressed earlier in this chapter, although there are high costs associated with climate action, there are even higher costs (regarding money, health, and safety) related to climate inaction.

The UN (2021a) wrote about some costs associated with climate change, stating a 1.8 trillion dollar global investment into adaptive strategies—such as systems that can provide warnings early on about natural disasters, constructing structures such as buildings and roads that can withstand various climates, and increase protection of sources of water—could end up being of great financial benefit. Although 1.8 trillion dollars is an incredibly large sum, they predicted that it would save 7.1 trillion dollars in the long term (UN, 2021a).

A 2019 article entitled *Financing Climate Change* by the UN (2020) states that before the pandemic, the World Bank estimated that by 2030, a 90 trillion dollar investment into infrastructure would be necessary for sustainability. Even with such a large number, the article writes that there are four times the benefits to costs—for every dollar spent, it would benefit by four dollars. The article also discussed a 2018 report that stated by 2030, the economy would benefit 26 trillion dollars if there was significant climate action (UN, 2020). It is important to note that the COVID-19 pandemic and various changes in the world could have changed these numbers, but they would likely still be in the trillions.

Furthermore, in an article released in 2022 by Deloitte—a firm that offers professional services related to finances, auditing, risk management, taxes, as well as many other services across the world—they discussed their research found that climate inaction would cost the global economy 178 trillion dollars by 2070 (Deloitte, 2022). Deloitte (2022) also suggests that if changes are effectively made, there is the potential for the global economy to benefit by 43 trillion dollars in the same time frame.

One article in *TIME* based on comments made at the United Nations Convention to Combat Desertification in Delhi, India in 2019 claimed that global warming could be stopped with 300 billion dollars (Majendie & Parija, 2019). The main argument discussed in the article was that by using the soil to lock in carbon, there could be significant steps made towards combating global warming. In large part, the argument brought up by scientists was about the damage to lands that became dried up, or in other words, desertification and that if carbon could be added to soil, the land could benefit (Majendie & Parija, 2019). This 300 billion dollar take, however, has been questioned. Reynolds (2019b) brought up some interesting points that must be considered when questioning if 300 billion dollars is an accurate sum of money to combat climate change. Notably, in his article *Can Soils Solve Climate Change?*, he brought up how there are no scientific papers (or at least none that are easily accessible to the public) that provide evidence to support the claims, but the articles are based solely on comments made at the United Nations Convention to Combat Desertification. The takeaway from this is that although approaching climate change from the perspective of increasing the amount of rich soil available, there are likely other important factors not being addressed or being overlooked completely. Moreover, it is important to understand that just because information is coming from professional sources, it does not mean that there are no gaps or limitations in the information being provided. For example, the conference (and subsequently the scientists attending) was focused on desertification, so the focus would have been primarily on strategies that address desertification-related issues. Reynolds (2019a) has questioned other claims, such as if it is accurate to claim that trees can solve climate change; based on a scientific article that stated planting trees was an incredible way to combat climate change. Reynolds (2019a; 2019b) sets a good example for individuals to look more critically at the scientific literature and conversations that are promoting various climate change strategies because he highlights how, although strategies may seem good on the surface, logistically and practically, there may be issues that make the strategies unattainable.

What Countries are Currently Doing about Climate Change

When looking into global efforts to combat climate change, it is important to know about the Paris Climate Agreement (also called the Paris Agreement). The Paris Climate Agreement is considered a "legally binding international treaty on climate change" (United Nations Framework Convention on Climate Change [UNFCCC], 2020). The focus of the treaty is to get the countries involved to reduce emissions in a timely manner and implement climate adaptation strategies (UN, 2021d). The agreement came about in 2015 and began being enforced in 2016, but there is still a long way to go for countries to achieve their emission-reduction goals (UNFCCC, 2020).

Looking into individual countries' actions, Canada has a 750 million dollar (in Canadian dollars) Emissions Reduction Fund that serves to help oil and gas companies adopt solutions to reducing greenhouse gas emissions (Government of Canada, 2022). Moreover, the United States has the Greenhouse Gas Reduction Fund which provides 27 billion dollars for the United States Environmental Protection Agency (EPA) to use for projects and strategies to reduce emissions (EPA, 2022). This is part of a larger commitment to the environment, called the Inflation Reduction Act which aims to reduce emissions by about 40 percent by the year 2030 (Senate Democrats, 2022). The act includes an investment of 369 billion dollars in Energy Security and Climate Change programs over 10 years (starting in 2022) (Senate Democrats, 2022). Around the world, countries have begun to implement strategies to adapt to climate change. The UN (2021a) writes about countries such as Peru, Costa Rica, Sri Lanka, Albania, and El Salvador and their respective ways of protecting and restoring their resources.

Although making large-scale global change is important, it is not always easy or possible for countries to implement effective strategies for climate adaptation and action. Sometimes there are barriers to climate action that must be considered when critically thinking about how governments should be taking action. For example, some countries have more finances available than others, meaning that some countries would likely have to aid others in financing climate action (UN, 2021a). Moreover, as previously discussed in the chapter, environments can become extremely unsafe, making relocation of individuals or populations more likely than implementing various strategies (UN, 2021a).

Putting Billions and Trillions of Dollars into Perspective

The amounts of money that are being discussed in relation to climate action and climate inaction are incredibly large. Trying to conceptualize and compare these numbers is incredibly difficult, so this section will provide examples and comparisons to help put the numbers into perspective. It must also be noted that the numbers discussed in this chapter must be taken with a grain of salt because creating estimates of sums of money this large that are impacted by several factors will not be exact. This has been made clear with the several different numbers estimated. The one thing, however, that all of the estimates have in common is that they are in the billions (or trillions) of dollars.

To start, to help visualize the large sums of money, one billion in digits is 1,000,000,000 (that is 9 zeros), while one trillion is 1,000,000,000,000 (that is 12 zeros). One thousand (1000) billion makes up one trillion. Looking at the UN's (2021a) prediction that climate action would cost 1.8 trillion dollars in comparison to 7.1 trillion dollars in benefits, that means that there would be 3.94 times the amount of benefits to costs. In other words, the benefits are approximately 25 percent greater than the costs.

Another comparison can be made when comparing the predicted costs of climate action and inaction on a global level to the gross domestic product (GDP) of various countries. This chapter will not delve into the details of what GDP represents and entails, but to provide a surface-level explanation, Callen (n.d.) explains GDP as "the monetary value of final goods and services ... produced in a country in a given period of time". As of 2022, the top 51 percent of global GDP comes from the United States, Japan, Germany, China, and India (Koop, 2022). Together, they amass approximately 55.1 trillion dollars; with the global GDP for 2022 being approximately 101.6 trillion dollars (Koop, 2022). Comparing some of the predicted values for how the global economy could benefit, Deloitte (2022) estimated that there would be a 178 trillion dollar cost to climate inaction by 2070. Averaged out, that would cost approximately 3.7 trillion dollars globally a year. 3.7 trillion dollars is higher than the majority of countries' GDPs—with only the United States, Japan, Germany, and China amassing higher in 2022 (Koop, 2022). This comparison serves only to show that the amount of money being discussed is extraordinarily large and requires a global united effort to accomplish. Seeing as the amounts necessary in a year are more than some countries' GDP, it is clear that in nations joining together, there is a much better chance of making significant changes.

Climate Change is a Relevant and Important Global Issue

As much as climate action is expensive, climate inaction will take a greater toll. Although nations around the world have started to commit to climate action, there is a long way to go before a real, beneficial change occurs. Climate change is currently impacting natural resources, natural disasters, and peace, amongst many other things. Only by investing globally will countries be able to save money in the long-term while also benefiting the health and safety of their citizens.

References

Al Gore. (n.d.). *The Climate Reality Project.* https://algore.com/project/the-climate-reality- project

Alter, C., Haynes, S., & Worland, J. (2019). *TIME 2019 person of the year | Great Thunberg.* TIME. https://time.com/person-of-the-year-2019-greta-thunberg/

Callen, T. (n.d.) *Gross domestic product: An economy's all.* International Monetary Fund. https://www.imf.org/en/Publications/fandd/issues/Series/Back-to-Basics/gross-domestic- product-GDP

David Suzuki Foundation. (n.d.). *Our Story.* https://davidsuzuki.org/about/our-story/

Deloitte. (2022, May 23). *Deloitte research reveals inaction on climate change could cost the world's economy US$178 trillion by 2070.* https://www.deloitte.com/global/en/about/press-room/deloitte-research-reveals- inaction-on-climate-change-could-cost-the-world-economy-us-dollar-178-trillion-by- 2070.html

Eldridge, A. (2022, March 20). *David Suzuki.* Encyclopedia Britannica. https://www.britannica.com/biography/David-Suzuki

Government of Canada. (December 15, 2022). *Emissions Reduction Fund: working together to create a lower carbon future.* https://www.nrcan.gc.ca/science-and-data/funding- partnerships/funding-opportunities/current-funding-opportunities/emissions-reduction-fund/22781

History.com Editors. (2022, August 8). *Climate Change History.* A&E Television Networks. https://www.history.com/topics/natural-disasters-and-environment/history-of-climate-change

Jackson, P. (2007, June). *From Stockholm to Kyoto: A brief history of climate change.* United   Nations. https://www.un.org/en/chronicle/article/stockholm-kyoto-brief-history- climate-change

Koop, A. (2022, December 29). *Top heavy: Countries by share of the global economy.* Visual Capitalist. https://www.visualcapitalist.com/countries-by-share-of-global-economy/

Majendie, A., & Parija, P. (2019, October 23). *These U.N. Climate Scientists Think They Can Halt Global Warming for $300 Billion. Here's How.* TIME. https://time.com/5709100/halt- climate-change-300-billion/

McFarland, J., Zhou, Y., Clarke, L., Sullivan, P., Colman, J., Jaglom, W. S., Colley, M., Patel, P., Eom, J., Kim, S. H., Kyle, G. P., Schultz, P., Venkatesh, B., Haydel, J., Mack, C., &   Creason, J. (2015). Impacts of rising air temperatures and emissions mitigation on electricity demand and supply in the United States: a multi-model comparison. *Climate Change, 131*, 111–125. https://link.springer.com/article/10.1007/s10584-015-1380-8

NobelPrize.org. (n.d.). *Al Gore – Facts.* https://www.nobelprize.org/prizes/peace/2007/gore/facts/

Reynolds, J. (2019a, July 9). *Can planting trees solve climate change?* Legal Planet. https://legal-planet.org/2019/07/05/can-planting-trees-solve-climate-change/#Update

Reynolds, J. (2019b, October 28). *Can soils solve climate change?* Legal Planet. https://legal-planet.org/2019/10/28/can-soils-solve-climate-change/

Senate Democrats. (2022). *Inflation reduction act one page summary.* *https://www.democrats.senate.gov/imo/media/doc/inflation_reduction_ act_one_page_su mmary.pdf*

The Climate Reality Project. (n.d.). [The Climate Reality Project main page]. https://www.climaterealityproject.org/

United Nations. (2021a, August 14). *Climate adaptation.* https://www. un.org/en/climatechange/climate-adaptation

United Nations. (2021b, November 1). *The Paris agreement.* https:// www.un.org/en/climatechange/paris-agreement

United Nations. (2020, December 6). *Financing climate action.* https:// www.un.org/en/climatechange/raising-ambition/climate-finance

United Nations. (2019, October 24). *The climate crisis – A race we can win.* https://www.un.org/en/un75/climate-crisis-race-we-can-win

United Nations Framework Convention on Climate Change. (2020, December 10). *The Paris agreement.* https://unfccc.int/process-and-meetings/the-paris-agreement/the-paris-agreement

United States Environmental Protection Agency. (2022, November 30). *Greenhouse gas reduction fund.* https://www.epa.gov/inflation-reduction-act/greenhouse-gas-reduction-fund

Chapter 10: Carbon Sinks, Or Sources?

Kelly Wu

Introduction

In the same way we inhale, hold, and exhale our breath, the Earth functions and "breathes" in the same cycles. This respiratory balance is of a delicate nature and works in tandem across several different types of biospheres and environments. With the occurrence of the Industrial Revolution and the consequences that followed, the exponential increase in human activity over the past few hundred years have thrown these natural rhythms off kilter. The burning of fossil fuels, pollution, and climate change have affected each step of this equilibrium, which as a result, has detrimental consequences for the future of human civilization. Wildfires are but a small but quite noticeable manifestation of this malignance. This chapter will give a brief overview of the importance of carbon, defining carbon sinks and sources, the carbon cycle, and the occurrence of wildfires in relation to these concepts, insofar that one can better understand their effects and how it, in return, contributes to climate change.

The Role of Carbon

Carbon, the sixth element, is foundational to organic matter given its unique property to create bonds with other atoms, such that it allows for a diverse range of biomolecules to form (Dohney-Adams, 2013). DNA and RNA, both which can mature and multiply, are one of such examples. When living creatures absorb carbon, whether that is through their lungs, gills, skin, food, or otherwise, they can use it towards energy that stimulates a variety of different cell functions, or it can be used to spur new cellular growth. As such, carbon is essential to all living things and is found all around us, from our very being to our food, water, rocks, soil, and air. The delicate balance formed in the ecosystems' exchange of carbon is important to the wellbeing of planet Earth, and can be affected by human interference and in turn, affect human wellbeing.

Defining Carbon Sinks, Sources, Cycle, and Sequestration

A carbon sink is an accumulation of carbon which intakes more of the element from the atmosphere than it puts out and can be either naturally occurring, such as the lithosphere, biosphere, or hydrosphere, or man made (Frape, 2016). Artificial carbon sinks may include specifically designed technology or chemical compounds to trap in carbon (Technology Academy Finland, 2022). Carbon sources are the opposite, which are origins for an excess output of carbon that outweighs the amount it absorbs (Client Earth, 2020). Carbon sources are more so associated with events such as volcanic eruptions, organic decomposition, the burning of fossil fuels, and, as the title of this book suggests, wildfires. The carbon cycle encompasses the way carbon moves through these global reservoirs, how it flows and is balanced in the sinks and sources across the world (National Geographic Society, 2022). One way in which carbon moves is through plants, which take in carbon dioxide from the atmosphere through photosynthesis. Organic matter die and decay into the Earth, dispersing the element into ground and soil, where bacteria can then return it back into the atmosphere by the process of decomposition. Lastly, carbon sequestration are techniques used to lock in carbon dioxide to prevent it from accumulating in the atmosphere (UC Davis, 2021). Through transforming the gaseous carbon into solid or liquid forms, this concept has been of particular interest to scientists over the past few decades given that it has significant applications in slowing down the effects of atmospheric global warming and climate change.

Types of Carbon Sinks

There is a diverse collection of carbon sinks that vary across geographical locations. However, this section will focus primarily on the main three relevant types, which are forests, soils, permafrost, and the ocean.

1) Forests

Earth's forests take around 7.6 million metric tonnes of carbon dioxide annually, which is approximately one and a half times more than what the United States releases per year (Harris & Gibbs, 2021). Coming in as the second largest carbon sink in the world, forests have helped moderate carbon release caused by human activities such as deforestation and burning of fossil fuel by taking in nearly one quarter of such emissions, with the industry knocking off around 38% of greenhouse gases during the ten year span between 2006 and 2016 (CCFM, 2022). Trees undergo photosynthesis, absorbing atmospheric carbon for energy and putting out oxygen, and then, in their burning or slow decay of their leaves or wood, release it back out as gas (UNECE, 2009). Norman & Kreye (2020) goes on to note the effect the maturity and tree density of a forest has on the speed of carbon sequestration, whereas younger woodlands experience more rapid growth (Norman & Kreye, 2020). However, as the trees compete for resources among the density, some may die out, releasing their captured carbon. Older forests with larger, more developed trees and roots structures mature slower but can sequester greater amounts, thus resulting in absorbing even larger volumes of emissions more effectively and efficiently.

However, forests cannot be relied on completely in the slowing of climate change. Recall that a carbon sink stores more emission than it creates. Due to the effects of human activity, the reverse is occurring, where these protective biospheres are instead becoming a contributor to global warming. Over the past two decades, approximately ten different World Heritage designated forests have become net carbon sources (Griffin, 2021). Canadian forests in particular have actually been a carbon source since 2001 due to invasive insects outbreaks and wildfires, the latter which accounted for a net sink of 237 megatonnes of emissions, that is, 237 megatonnes of more CO_2 released than absorbed in 2015 (Fletcher, 2019). Frequency of wildfires have also increased in Canada over the years, with the province of British Columbia seeing a massive woodland emissions jump in both 2017 and 2018 that was triple of the cumulative carbon discharge from all sources in 2016 (Government of Canada, 2022; Jones, 2019). Factoring 43 million tonnes from logging, 4 million from slash burning (a type of forestry management technique where land is cut down and burnt away to clear out a field in order to increase soil fertility), and another 190 million from wildfires, that is a total of 190 million tonnes of emissions that was unaccounted for by the Canadian government in 2017. (Allen, 2021; Wieting, 2019)

In terms of forests on a global scale, a 2020 study found that tropical forests' ability to store carbon are diminishing, lessening their role as a carbon sink (Hubau et al., 2020). According to Hubau et al (2020), the impact from the logging and farming industry, droughts, and deforestation has caused a drop in one third of carbon absorbed compared to the amount absorbed in 1990. In a worrying turn of events, it is predicted that the Amazon forest will soon make the switch from sink to source in the next decade or so, putting humanity's emissions years beyond what was expected of the most discouraging statistical models (Harvey, 2020).

2) Soils

A mixture between minerals, organic matter, gases, liquids, and living organisms, soil makes up about 10% of Earth's surface (Queensland Government, 2013; TERC, 2010). Of such ingredients in particular, it is composed of deteriorating plant matter, which in and of itself contains the carbon it absorbed. Soils, as well as the plants thriving in it, take in around one third of carbon emissions (Carrington, 2021). An article by Melillo & Gribkoff (2021) notes that the lower the temperature, the longer the soil can sequester, as the rate of decay and decomposition slows, a factor that plays into the chain reaction of climate change and the carbon cycle. The text further notes that in addition to agricultural lands and specific planted crops, scientists estimate this specific carbon sink is holding back approximately one billion tons of carbon per year. Perennial plants, or recurring plants that die in the winter and regrow from their rootstock the following spring, are one of such crops (Iannotti, 2021). Enduring vegetation such as fruit trees, oil palms, and certain berries have far extending roots who's seedlings are reborn from and are more effective in trapping carbon in the ground over a longer period of time compared to annual plants (PFAF, 2020).

Soil has long been cited as one of the largest carbon sinks, which can sequester emissions for hundreds of years compared to biomass, which decay and release the element quickly after death (Carrington, 2021; Client Earth, 2020). Results from recent experiments however, have come to challenge that claim, and its role may be smaller than what was previously thought. Terrer et. al (2020) analyzed 180 experiments ranging from different geological soils and vegetation that were put under excess carbon dioxide levels, finding that though the amount that plants and trees sequestered increased, the soils were not able to increase its ability to store it in the same proportion. The study hypothesized an inverse relation between the two, as they saw an increase in plant biomass but stagnant, or even occasionally, declining SOC (soil organic carbon) intake under high carbon dioxide conditions. The authors cited the increased plant growth to require more nutrition, which encourages root stimulation, a response that actually causes additional carbon dioxide soil release, as a potential explanation to their findings (Carrington, 2021). Overall, the conclusion to this paper may imply an overreliance on soils' role in the fight against global warming and may require a restructured approach in our efforts to deter climate change.

Wildfires' effect on this particular biosphere should not be understated. Though forests seem to be the most likely to be gravely affected, the spike in frequency of these disasters hinder carbon retention. With each fire event, layers beneath the topsoil avoid being scorched, leading to a buildup of organic carbon soil, so called 'legacy carbon', which can sequester the element for centuries (Morrison, 2019). According to Walker et. al (2019), as boreal wildfires heighten in magnitude, prevalence, and range due to climate change, the burning of this deep, carbon rich organic soil will release excess emissions, contributing to soils' shift from sink to source (Walker et al., 2019).

3) Permafrost

Maintaining a temperature of 0°C or lower, permafrost is a ground layer of the Earth composed of soil, gravel, and sand that is held together by ice (Rutledge et al., 2022). This type of sink contains over 1,600 billion tonnes of carbon, which is double the extent to which is stored in the atmosphere (Brouillette, 2021). As another one of Earth's historically significant carbon sinks, it can cover whole entire biomes, such as across the Arctic tundra, or smaller regions, such as mountain peaks, its overall reach ranging over 21 million kilometers squared, or about 11% of Earth's surface (Obu, 2021). Like all other aspects of the environment however, rising temperatures are greatly damaging these natural stores. According to the National Oceanic and Atmospheric Administration, arctic amplification, the phenomenon where arctic climates specifically are climbing in a higher proportion to the rest of the world, is developing at a rate two times higher than the global average since 2000 (Scott, 2020). Worryingly enough, it is estimated that the upper layers of permafrost could recede up to two-thirds by the year of 2100 (European Space Agency, 2021). When frozen, organic biomass locked in the ice does not decay and release carbon. Its thawing allows for century-old bacteria and microorganisms that may be immune to modern day medicine to escape, with global warming posing not only an environmental risk but a biohazardous one as well (Cho, 2022). Over the 14 years across 2003 to 2017, approximately 1.7 billion metric tons of carbon was released from the Arctic, with scientists estimating its thawing discharging more emissions than the cumulative amount created by human fossil fuels activity (NASA, 2019).

Wildfires' impact on permafrost is particularly noticeable in the regions near the Arctic Circle. A study by Yanagiya & Furuya (2020) inquired on the permafrost status in Siberia after several 2014 blazes that tore away approximately 3.5 million cubic meters of permafrost. The study noted that while the fire does not directly melt the permafrost, it incinerates the top flora which acts as a coolant as it absorbs and reflects daylight, which contributes to dehydration (Kornei, 2020).

Compared to forests, Holloway et al (2020) noted that specific lowland forests and tundra regions fare better against these infernos, returning to their original soil density over time, whereas uplands and environments with dry and lean organic soil will be more significantly and permanently altered (Holloway et al, 2020). UNESCO World Heritage Site Lena Pillars, a Siberian natural park with primary boreal vegetation, is one of the world's coldest and most isolated regions, had an outbreak of over 300 fires in the summer of 2021 (The Siberian Times, 2021; World Heritage Datasheet, 2017). In Canada, roughly 25,000 km squared of boreal land is burned in Canada, where a majority of the region is composed of forested peatland (Li et al., 2021). Peatlands soil contains an abundance of carbon due to the damp conditions that greatly slow organic decay (International Peatland Society, 2020). The waterlogged plants can take over thousands of years to decompose, allowing a northern peatland to accumulate almost ten times more carbon as a boreal forest (Witze, 2020). The article further notes that unlike woodlands, peatlands are more at risk as they do not recover the carbon lost as fast as tree regrowth. Worryingly enough, Smith (2020) estimates that half of Arctic wildfires during May and June 2020 affected peatlands, with several lasting the course of many days, potentially spurred on by the dense natural material (Smith, 2020).

4) Oceans

Oceans play a crucial role in regulating the Earth's climate, storing around 31% of atmospheric carbon dioxide (NCEI, 2022). In that way, as gaseous CO_2 increases, so does the amount in the water in a process known as ocean acidification (NOAA, 2020). Over 525 billion tons of emissions have been absorbed since the Industrial Revolution, a present rate which has stabilized to approximately 22 million tons a day, in a phenomenon that affects the pH scale of the water (Bennett, 2019). As carbon dioxide is absorbed by the oceans, it reacts with the water to form carbonic acid, which can then be neutralized by the oceans' natural buffers (EPA, 2016). The consequences of this excess element in the water is detrimental to ocean life as it reduces the concentration of carbonate, a molecule that is foundational for shell animals and coral reefs to survive (Ireland & Hu, 2022). Increased water acidity could quite literally dissolve shells, which would affect the marine life food chain and contribute to species endangerment (Natural History Museum, 2019). In regards to coral reefs, a home to almost 25% of all known aquatic life, ocean acidification weakens their natural growth (Florida Keys National Marine Sanctuary, 2011; National Science Foundation, 2011). Moving from a carbon sink to a potential source, the decomposition of organic matter at the bottom of the ocean acts as a point of emission as it releases carbon dioxide into the water, a process that can be exacerbated by warming waters (Riebeek, 2011).

It may seem antithetical to draw a relation between wildfires and oceans but as reiterated before, the balance between each stage of the carbon cycle is delicate and deeply interrelated. In this regard, carbon emissions from wildfires can indeed affect aquatic life and wellbeing of the waters. One third of carbon from wildfires ends up in the ocean, where it resides there significantly longer than it does on land, over a period of up to many millenia (Vaughan, 2020)! For example, the devastating 2019 Australian wildfire that torched 21% of the country's forests contained iron in their smoke, which ended up spreading thousands of kilometers out into the water and essentially acting as fertilizer for marine vegetation (Jones, 2021). Two years later, algae growth erupted off the coast, blooming abnormally to almost the width of Australia itself at 2,000 miles long!

The Carbon Cycle Today

Alterations to one sink or source can cause a domino chain of developments affecting all aspects of the carbon cycle. The Industrial Revolution and beginning of humanity's dependence on fossil fuels have greatly accelerated the amount of carbon released into the atmosphere, with molecular concentrations increasing 39%, from 280 parts per million to 387 parts per million (NASA, 2011). The current concentration is the highest it has been in the past 3.6 million years (NOAA, 2019)! In the past, the role of natural forest wildfires was to have a necessary release of greenhouse gasses and carbon through the incineration of the organic matter consumed, with the remaining material either dying or undergoing a more hasty decay, which would further release carbon (California Department of Forestry and Fire Protection, 2022). In its regrowth, it draws back atmospheric carbon, cools temperatures with the vegetation reflecting sunlight, and the residual charcoal acts as a carbon sink by decomposing slower (up to several hundred years) and storing more carbon, making up 12% of the carbon released in these inferno events (Swansea University, 2020). However, factoring in all other contributors to climate change, this cycle of fire is becoming more intense and frequent. For example, heightened global temperatures, droughts, and dry environments contribute to the dehydration of vegetation, increasing flammability, which has been cited to be the reason why fires between 1984 and 2015 have doubled in the United States (Center for Climate and Energy Solutions, 2022). The rising heat also shortens winter and snow season by around one month, prolonging the dry season (EDF, 2018). In most Western regions in America, an approximate jump of 1 degree celsius per year would result in a 600% affected area increase (NOAA, 2022).

Conclusion

Carbon is a vital element that plays a key role in the functioning of the Earth's climate system. As a building block of life, carbon is found in all living things and is essential for the proper functioning of many biological processes. Carbon sinks, such as forests, soil, permafrost, and oceans absorb and store carbon dioxide from the atmosphere, helping to mitigate the greenhouse effect and global warming. Yet with growing threats of climate change, these protective assets are shifting to becoming carbon sources. Wildfires can have a significant impact on this delicate cycle and can lead to a feedback loop, as the warming caused by the increase in atmospheric carbon dioxide can then lead to more wildfires and deforestation. By understanding the importance of carbon and the role of carbon sinks and sources in the carbon cycle, we can take action to protect the Earth's climate and preserve the natural world for future generations.

References

Allen, S. (2021). Slash and burn: Learning to farm sustainably after Indonesia's wildfires. FED FED FED. Retrieved December 12, 2022, from https://fedfedfed.com/sliced/slash-and-burn#:~:text=and%20 profit%20margins.-,Slash%20and%20burn%20is%20a%20 12%2C000%20year%2Dold%20farming%20technique,during%20 the%20next%20rainy%20season.

Bennett, J. (2019, June 20). Ocean acidification. Smithsonian Ocean. Retrieved December 15, 2022, from https://ocean.si.edu/ocean-life/invertebrates/ocean-acidification

Brouillette, M. (2021, March 17). How microbes in permafrost could trigger a massive carbon bomb. Nature News. Retrieved December 12, 2022, from https://www.nature.com/articles/d41586-021-00659-y

CCFM. (2021, March 20). Forests: A stabilizing force for the climate. Canadian Council of Forest Ministers (CCFM). Retrieved December 12, 2022, from https://www.ccfm.org/climate-conscious/forests-a-stabilizing-force-for-the-climate/

California Department of Forestry and Fire Protection. (2022). Forests have an important role in climate solutions. Cal Fire Department of Forestry and Fire Protection. Retrieved December 15, 2022, from https://www.fire.ca.gov/programs/resource-management/resource-protection-improvement/wildfire-resilience/forest-stewardship/carbon-sequestration-and-a-changing-climate/#:~:text=Fire%20is%20an%20important%20part,and%20decompose%20rapidly%2C%20releasing%20CO2

Carrington, D. (2021, March 24). One of Earth's giant carbon sinks may have been overestimated - study. The Guardian. Retrieved December 12, 2022, from https://www.theguardian.com/environment/2021/mar/24/soils-ability-to-absorb-carbon-emissions-may-be-overestimated-study

Center for Climate and Energy Solutions. (2022, May 18). Wildfires and climate change. Center for Climate and Energy Solutions. Retrieved December 15, 2022, from https://www.c2es.org/content/wildfires-and-climate-change/

Cho, R. (2022, September 13). What lies beneath melting glaciers and thawing permafrost? State of the Planet. Retrieved December 15, 2022, from https://news.climate.columbia.edu/2022/09/13/what-lies-beneath-melting-glaciers-and-thawing-permafrost/#:~:text=One%20gram%20of%20permafrost%20was,as%20smallpox%20or%20Bubonic%20plague

Client Earth. (2020). What is a carbon sink? ClientEarth. Retrieved December 12, 2022, from https://www.clientearth.org/latest/latest-updates/stories/what-is-a-carbon-sink/#:~:text=Examples%20of%20carbon%20sources%20include,natural%20carbon%20sinks%20can%20absorb.

Doheny-Adams, T. (2013). The importance of carbon to life. FutureLearn. Retrieved December 12, 2022, from https://www.futurelearn.com/info/courses/the-biology-of-bugs-brains-and-beasts/0/steps/68848

EDF. (2018, July 19). Here's how climate change affects wildfires. Environmental Defense Fund. Retrieved December 15, 2022, from https://www.edf.org/climate/heres-how-climate-change-affects-wildfires

EPA. (2016). Understanding the Science of Ocean and Coastal Acidification. EPA. Retrieved December 15, 2022, from https://www.epa.gov/ocean-acidification/understanding-science-ocean-and-coastal-acidification

European Space Agency. (2021, October 22). Permafrost thaw could release bacteria and viruses. ESA. Retrieved December 15, 2022, from https://www.esa.int/Applications/Observing_the_Earth/Permafrost_thaw_could_release_bacteria_and_viruses

Fletcher, R. (2019, February 12). Canada's forests actually emit more carbon than they absorb — despite what you've heard on Facebook. CBCnews. Retrieved December 12, 2022, from https://www.cbc.ca/news/canada/calgary/canada-forests-carbon-sink-or-source-1.5011490

Florida Keys National Marine Sanctuary. (2011, April 4). Coral reefs are massive structures made of limestone deposited by coral polyps. What is a coral reef? Retrieved December 15, 2022, from https://floridakeys. noaa.gov/corals/coralreef.html

Frape, D. (2016). The functions and sizes of the five carbon sinks on Planet Earth and their relation to climate change part I their present sizes and locations. World Agriculture. Retrieved December 12, 2022, from http://www.world-agriculture.net/article/the-functions-and-sizes-of-the-five-carbon-sinks-on-planet-earth-and-their-relation-to-climate-change-part-i-their-present-sizes

Government of Canada. (2022, August 11). Government of Canada. Retrieved December 12, 2022, from https://www.nrcan.gc.ca/our-natural-resources/forests/wildland-fires-insects-disturbances/climate-change-fire/13155

Griffin, O. (2021, October 28). Wildfires, logging turn protected forests into carbon emitters -report. Reuters. Retrieved December 12, 2022, from https://www.reuters.com/business/cop/wildfires-logging-turn-protected-forests-into-carbon-emitters-report-2021-10-27/

Harris, N., & Gibbs, D. (2021, January 21). Forests absorb twice as much carbon as they emit each year. World Resources Institute. Retrieved December 12, 2022, from https://www.wri.org/insights/forests-absorb-twice-much-carbon-they-emit-each-year

Harvey, F. (2020, March 4). Tropical forests losing their ability to absorb carbon, study finds. The Guardian. Retrieved December 15, 2022, from https://www.theguardian.com/environment/2020/mar/04/tropical-forests-losing-their-ability-to-absorb-carbon-study-finds

Holloway, J. E., Lewkowicz, A. G., Douglas, T. A., Li, X., Turetsky, M. R., Baltzer, J. L., & Jin, H. (2020). Impact of wildfire on Permafrost Landscapes: A review of recent advances and future prospects. Permafrost and Periglacial Processes, 31(3), 371–382. https://doi.org/10.1002/ppp.2048

Hubau, W., Lewis, S. L., Phillips, O. L., Affum-Baffoe, K., Beeckman, H., Cuní-Sanchez, A., Daniels, A. K., Ewango, C. E. N., Fauset, S., Mukinzi, J. M., Sheil, D., Sonké, B., Sullivan, M. J. P., Sunderland, T. C. H., Taedoumg, H., Thomas, S. C., White, L. J. T., Abernethy, K. A., Adu-Bredu, S., … Zemagho, L. (2020, March 4). Asynchronous carbon sink saturation in African and Amazonian tropical forests. Nature News. Retrieved December 15, 2022, from https://www.nature.com/articles/s41586-020-2035-0

Iannotti, M. (2021, December 9). What is perennial and what are the different types? The Spruce. Retrieved December 12, 2022, from https://www.thespruce.com/what-is-a-perennial-flower-or-plant-1402789

International Peatland Society. (2020, September 22). What are peatlands? International Peatland Society. Retrieved December 15, 2022, from https://peatlands.org/peatlands/what-are-peatlands/

Ireland, P., & Hu, S. (2022, October 13). Ocean acidification: What you need to know. NRDC. Retrieved December 15, 2022, from https://www.nrdc.org/stories/what-you-need-know-about-ocean-acidification#people

Jones, B. (2021, September 15). Wildfires in Australia caused an explosion of sea life thousands of miles away. Vox. Retrieved December 15, 2022, from https://www.vox.com/down-to-earth/2021/9/15/22672480/wildfires-oceans-algae-blooms-climate-change-australia

Jones, R. P. (2019, January 28). B.C. forests contribute 'hidden' carbon emissions that dwarf official numbers, report says. CBCnews. Retrieved December 12, 2022, from https://www.cbc.ca/news/canada/british-columbia/sierra-club-report-forest-carbon-emissions-1.4995191

Kornei, K. (2020). Wildfires trigger long-term permafrost thawing. Eos, 101. https://doi.org/10.1029/2020eo148336

Li, X.-Y., Jin, H.-J., Wang, H.-W., Marchenko, S. S., Shan, W., Luo, D.-L., He, R.-X., Spektor, V., Huang, Y.-D., Li, X.-Y., & Jia, N. (2021). Influences of forest fires on the Permafrost Environment: A Review. Advances in Climate Change Research, 12(1), 48–65. https://doi.org/10.1016/j.accre.2021.01.001

Melillo, J., & Gribkoff, E. (2021, April 15). Soil-based carbon sequestration. MIT Climate Portal. Retrieved December 12, 2022, from https://climate.mit.edu/explainers/soil-based-carbon-sequestration

Morrison, L. C. (2019, August 21). Legacy carbon being released by more extreme wildfires, Canadian-American team reports in Nature. University of Laurier. Retrieved December 12, 2022, from https://www.wlu.ca/news/news-releases/2019/aug/legacy-carbon-being-released-by-more-extreme-wildfires-canadian-american-team-reports-in-nature.html

NASA. (2011, June 16). Changes in the Carbon Cycle. NASA. Retrieved December 15, 2022, from https://earthobservatory.nasa.gov/features/CarbonCycle/page4.php

NASA. (2019, November 18). Permafrost becoming a carbon source instead of a sink. NASA. Retrieved December 15, 2022, from https://earthobservatory.nasa.gov/images/145880/permafrost-becoming-a-carbon-source-instead-of-a-sink

NCEI. (2022, August 26). Quantifying the Ocean Carbon Sink. National Centers for Environmental Information (NCEI). Retrieved December 15, 2022, from https://www.ncei.noaa.gov/news/quantifying-ocean-carbon-sink#:~:text=The%20ocean%20acts%20as%20a,2%20levels%20in%20the%20ocean.

NOAA. (2019, February 1). Carbon cycle. National Oceanic and Atmospheric Administration. Retrieved December 15, 2022, from https://www.noaa.gov/education/resource-collections/climate/carbon-cycle#:~:text=Carbon%20is%20the%20chemical%20backbone,that%20fuels%20our%20global%20economy.&text=Most%20of%20Earth's%20carbon%20is,atmosphere%2C%20and%20in%20living%20organisms

NOAA. (2020, April 1). Ocean acidification. National Oceanic and Atmospheric Administration. Retrieved December 15, 2022, from https://www.noaa.gov/education/resource-collections/ocean-coasts/ocean-acidification

NOAA. (2022, August 8). Wildfire climate connection. National Oceanic and Atmospheric Administration. Retrieved December 15, 2022, from https://www.noaa.gov/noaa-wildfire/wildfire-climate-connection

National Geographic Society. (2022). Carbon sources and sinks. National Geographic Society. Retrieved December 12, 2022, from https://education.nationalgeographic.org/resource/carbon-sources-and-sinks

National Science Foundation. (2018, January 29). Scientists pinpoint how ocean acidification weakens coral skeletons. NSF. Retrieved December 15, 2022, from https://beta.nsf.gov/news/scientists-pinpoint-how-ocean-acidification#:~:text=The%20rising%20acidity%20of%20the,corals%20will%20be%20most%20vulnerable

Natural History Museum, H. (2019, April 30). How does ocean acidification affect marine life? Retrieved December 15, 2022, from https://www.nhm.ac.uk/discover/quick-questions/how-does-ocean-acidification-affect-marine-life.html#:~:text=Ocean%20acidification%20can%20negatively%20affect,the%20faster%20the%20shells%20dissolve

Norman, C., & Kreye, M. (2020). How forests store carbon. Penn State Extension. Retrieved December 12, 2022, from https://extension.psu.edu/how-forests-store-carbon

Obu, J. (2021). How much of the Earth's surface is underlain by permafrost? Journal of Geophysical Research: Earth Surface, 126(5). https://doi.org/10.1029/2021jf006123

PFAF. (2020, August 21). Carbon Sequestration. Plants for a Future. Retrieved December 12, 2022, from https://pfaf.org/user/cmspage.aspx?pageid=324

Queensland Government. (2013, October 8). How soils form. Queensland Government. Retrieved December 12, 2022, from https://www.qld.gov.au/environment/land/management/soil/soil-explained/forms#:~:text=Soil%20is%20the%20thin%20layer,which%20interact%20slowly%20yet%20constantly.

Riebeek, H. (2011, June 16). The carbon cycle. NASA. Retrieved December 15, 2022, from https://earthobservatory.nasa.gov/features/CarbonCycle

Rutledge, K., McDaniel, L., Teng, S., Hall, H., Ramroop, T., Sprout, E., Hunt, J., Boudreau, D., & Costa, H. I. (2022, May 20). Permafrost. National Geographic Society. Retrieved December 12, 2022, from https://education.nationalgeographic.org/resource/permafrost

Scott, M. (2020, December 8). 2020 arctic air temperatures continue a long-term warming streak. NOAA Climate.gov. Retrieved December 15, 2022, from https://www.climate.gov/news-features/featured-images/2020-arctic-air-temperatures-continue-long-term-warming-streak

Smith, T. (2020, July 21). New spatial analysis of wildfires across the Arctic in May/June 2020, and how they compare to the satellite record (2003-2020). What is burning? Are there peat fires? What about permafrost? New spatial analysis of wildfires across the Arctic in May/June 2020, and how they compare to the satellite record (2003-2020). What is burn... Retrieved December 15, 2022, from https://threadreaderapp.com/thread/1285514278527737858.html

Swansea University. (2020, February 12). Home. Swansea University. Retrieved December 15, 2022, from https://www.swansea.ac.uk/research/research-highlights/sustainable-futures-energy-environment/wildfires/

TERC. (2010, November 22). Part A: Earth System Science. Earth System Science. Retrieved December 12, 2022, from https://serc.carleton.edu/eslabs/climate/1a.html#:~:text=Although%20soil%20only%20accounts%20for,the%20atmosphere%20to%20form%20rain

Technology Academy Finland. (2022, July 20). Artificial Carbon Sinks explained. Millennium Technology Prize. Retrieved December 12, 2022, from https://millenniumprize.org/news-articles/news/artificial-carbon-sinks-explained/

Terrer, C., Phillips, R. P., Hungate, B. A., Rosende, J., Pett-Ridge, J., Craig, M. E., van Groenigen, K. J., Keenan, T. F., Sulman, B. N., Stocker, B. D., Reich, P. B., Pellegrini, A. F., Pendall, E., Zhang, H., Evans, R. D., Carrillo, Y., Fisher, J. B., Van Sundert, K., Vicca, S., & Jackson, R. B. (2021). A trade-off between plant and soil carbon storage under elevated CO2. Nature, 591(7851), 599–603. https://doi.org/10.1038/s41586-021-03306-8

The Siberian Times. (2021, July 13). Permafrost is ablaze with hundreds of wildfires in world's coldest region. The Siberian Times. Retrieved December 15, 2022, from https://siberiantimes.com/other/others/features/permafrost-is-ablaze-with-hundreds-of-wildfires-in-worlds-coldest-region/

UC Davis. (2021, January 31). Carbon sequestration. UC Davis. Retrieved December 12, 2022, from https://climatechange.ucdavis.edu/climate/definitions/carbon-sequestration

UNECE. (2009). Carbon Sinks and Sequestration. UNECE. Retrieved December 12, 2022, from https://unece.org/forests/carbon-sinks-and-sequestration

Vaughan, A. (2020, June 18). Huge amounts of carbon from forest fires ends up in the Ocean. New Scientist. Retrieved December 15, 2022, from https://www.newscientist.com/article/2245038-huge-amounts-of-carbon-from-forest-fires-ends-up-in-the-ocean/

Walker, X. J., Baltzer, J. L., Cumming, S. G., Day, N. J., Ebert, C., Goetz, S., Johnstone, J. F., Potter, S., Rogers, B. M., Schuur, E. A., Turetsky, M. R., & Mack, M. C. (2019). Increasing wildfires threaten historic carbon sink of boreal forest soils. Nature, 572(7770), 520–523. https://doi.org/10.1038/s41586-019-1474-y

Wieting, J. (2019, January). Sierra Club BC. Retrieved December 12, 2022, from https://sierraclub.bc.ca/

Witze, A. (2020). The Arctic is burning like never before — and that's bad news for climate change. Nature, 585(7825), 336–337. https://doi.org/10.1038/d41586-020-02568-y

World Heritage Datasheet. (2017, May 22). Lena Pillars Nature Park. World Heritage Datasheet. Retrieved December 15, 2022, from http://world-heritage-datasheets.unep-wcmc.org/datasheet/output/site/lena-pillars-nature-park/

Chapter 11: An Unwelcome Visitor

Shanuga Thavarajah

Introduction

The effects of wildfires on animals are twofold, there are the effects as a result of the fire itself and the effects from the aftermath of the fire. While many animals in fire-prone habitats possess wildfire detection and avoidance behaviors, not all animals are able to escape the effects of the fire and experience physical injuries such as burns and injuries from smoke inhalation (Sanderfoot et al., 2021). There are also the effects of habitat loss that affects an animal's access to food and other resources ultimately resulting in potential population declines (ifaw, 2021). Throughout this chapter, the different impacts wildfires have on animals will be examined in closer detail.

Physical Injuries from Wildfires

Burn Injuries

The most commonly seen impact of wildfires on wildlife is burn injuries from the fires. These burns may occur as a result of the flames of the fire themselves or from the heat of the fire (Butkus et al., 2021). While wildfires have been increasing in number over the past few decades, research on the impact they have in terms of direct effects on wildlife populations is lacking (Butkus et al., 2021). As a result of this deficiency in data and research, there is a lack of knowledge of potential treatment options for animals suffering from burn injuries from wildfires

(Butkus et al., 2021). An online survey administered to international wildlife rehabilitation facilities by Butkus and team aimed to find baseline findings on the number of burn cases admitted, treatments used and survivorship of wildlife following burns injuries (2021). The purpose of collecting these findings is to examine and identify areas to implement any potential changes in burn care as well as identify areas of improvement for wildlife care (Butkus et al., 2021). The study found that approximately 80% of admissions to the surveyed wildlife facilities were for wildfire burns from the years of 2015 to 2018 (Butkus et al., 2021). Bandages, colloid fluids and opioids were commonly used in facilities that had veterinarians, although animal survival did not differ between facilities that had or did not have veterinarians (Butkus et al., 2021). The survey results also indicate that 88% of facilities reported animal scarring, 81% reported wildlife alopecia and 61% reported sepsis from the burns (Butkus et al., 2021). It was also reported that wildlife admitted to the facilities had equal odds of surviving and dying (Butkus et al., 2021). With these initial findings, it is important to conduct more research on burn injury treatments to increase the survival of animals experiencing burn injuries.

Smoke Inhalation on Animal Populations

There are not many studies that examine the effect that smoke inhalation from wildfires has on animal populations and those that do are typically associated with other disciplines, with this relationship not the primary purpose of those studies (Sanderfoot et al., 2021). Considering the frequency and magnitude of the wildfires that have occurred within the last few decades, research examining the effects of smoke from wildfires on animals is crucial as it can help us better treat these animals following a wildfire. Additionally, wildfire smoke is capable of large distribution, affecting surrounding areas that may not have been impacted by the wildfire. In these cases, animals may be facing the effects of smoke inhalation without having faced the wildfire, and understanding how wildfire smoke affects animals can lead to faster diagnosis and treatment of these animals.

In a review of 41 studies, it was concluded that wildfire smoke contributes to adverse acute and chronic health outcomes (Sanderfoot et al., 2021). To further expand, smoke inhalation was implicated in carbon monoxide poisoning, respiratory distress, neurological impairment, respiratory and cardiovascular disease, oxidative stress, and immunosuppression, and these findings were seen in both terrestrial and aquatic organisms (Sanderfoot et al., 2021). First and foremost, smoke inhalation results in thermal and chemical damage to the lung tissue within vertebrate species (Sanderfoot et al., 2021). From this, there can be fluid accumulation within the lungs, also known as pulmonary edema (Sanderfoot et al., 2021). This has been documented in livestock and pets when structural fires such as house fires, and considering that these fires are typically smaller on a scale compared to a forest burning, there must be a greater effect on animals from a forest wildfire (Sanderfoot et al., 2021). Laboured breathing, rapid breathing, panting, foaming at the nostrils, wheezing, and a rapid heart rate are all symptoms of respiratory distress syndrome, which is also seen in animals following smoke inhalation (Sanderfoot et al., 2021). If untreated, the animals will experience hypoxemia as a result of impairments to gas exchange and acidosis (elevated acid levels in the blood) (Sanderfoot et al., 2021). This is seen in captive bottlenose dolphins following a fire in 2003 as carbon dioxide levels had elevated following the fire (Venn-Watson et al., 2013).

There are also detriments to the immune system. Within mammals, smoke inhalation causes the production of macrophages and lymphocytes, but there is an alteration to their function as a result of the smoke such that the immune cells will not function improperly (Sanderfoot et al., 2021). Toxins found within the smoke are also capable of destroying antioxidants such that the body is unable to remove free radicals, resulting in oxidative stress (Sanderfoot et al., 2021). Moreover, this increase in oxidative stress can further impair immune system function. Ultimately, the animal is even more vulnerable to other diseases affecting them due to being immunocompromised. Take the captive bottlenose dolphins that had acidosis following the wildfire (Venn-Watson et al., 2013). Along with the acidosis, the dolphins were also found to be more likely to contract pneumonia following the wildfire than normally (Venn-Watson et al., 2013).

Smoke contains several toxins such as carbon monoxide, hydrogen cyanide, and particulate matter. Inhalation of these toxins by animals can be fatal, especially carbon monoxide (Sanderfoot et al., 2021). With carbon monoxide poisoning, the carbon monoxide will bind to hemoglobin, preventing oxygen from binding to hemoglobin (Sanderfoot et al., 2021). Hemoglobin is the molecule responsible for distributing oxygen systemically and with low levels of oxygen being transported, organs and tissues within the body experience hypoxia (insufficient oxygen) (Sanderfoot et al., 2021). Moreover, prolonged hypoxia can irreversibly damage organs and tissues and even affect the fleeing ability of animals during wildfires as hypoxia is capable of causing neurological damage such as disorientation.

Smoke inhalation can also cause shifts in animal behavior. It has been hypothesized that these changes may be due to health effects from smoke inhalation or protective behaviors against further smoke inhalation (Sanderfoot et al., 2021). In terms of changes in behavior, animals may choose to conserve their energy in response to wildfires and the smoke. In addition, animals may get more agitated as a result of the smoke (Sanderfoot et al., 2021). For instance, pets and livestock become agitated when exposed to smoke from structural fires (Sanderfoot et al., 2021). Along with their agitation, they also exhibit more vocalization, less physical activity and neurological impairments such as disorientation (Sanderfoot et al., 2021). Finally, there is research showing that the vegetation that is burned along with the stage and severity of the fire as well as the distance the smoke can travel all factor into how the smoke affects animal populations (Sanderfoot et al., 2021).

Habitat Loss from Wildfires

Damage to Terrestrial Organisms

One of the major impacts caused by a wildfire is habitat loss which impacts the food, water, shelter and other resources an animal may find within that habitat (ifaw, 2021). The lack of these resources increases competition and ultimately reduces the size of certain animal populations (ifaw, 2021). The wildfire may even cause a habitat to be more suitable for invasive species, further increasing the competition the native species may face. Ultimately, habitat loss from a wildfire can result in species loss, and should the species be endemic to that area, it may even result in the extinction of that species. Take for example the biodiversity loss within Central Chile as a result of human-induced wildfires (Braun et al., 2021). Central Chile is a biodiversity hotspot, containing a high number of species and the species within this area are threatened as a result of anthropogenic actions (Braun et al., 2021). From this study, Braun et al., conclude that the impacts of wildfires along with land use changes increase the rate of biodiversity loss within this ecosystem (2021).

To mitigate and control biodiversity loss as a result of wildfires, conservation biologists have developed a model known as the Species Distribution Model (SDM) (Bossoe et al., 2018). This model is used to understand how species respond to wildfires and a 2018 study using this model shows its efficacy (Bossoe et al., 2018). Following the 2017 wildfire in Italy, the SDM model was used to gauge the habitat loss of 12 of the bat species residing within this area (Bossoe et al., 2018). The model supported the initial hypothesis of the researchers that the fire did affect all species as a result of habitat loss and that the bats that forage the most were the ones affected the most (Bossoe et al., 2018). Hence, SDM is a useful tool for assessing the aftermath of wildfires in a quick manner. Going further, the use of the model shows which areas within a landscape are hotspots for animals and in terms of protecting biodiversity, these areas can be protected such that in the event of a wildfire, these animals are little to or not affected at all (Bossoe et al., 2018).

However, to say that wildfires result in the decrease of all species would be an inaccurate statement as wildfires can have different effects on species based on their habitat use and the severity of the fire. From fire disturbances, there can be an increase, decrease or even a curvilinear relationship in species abundance (Lewis et al., 2022). The relationship between the species abundance in response to fire disturbances is based on traits such as life history characteristics which includes fecundity, as well as habitat association and use of the animal. Fire severity is described by the number of fire burns occurring on a gradient. An unburned forest is where fire has not occurred for an extended period of time, low fire severity is where the fire burns the understory but not the mature trees, and moderate fire severity is where the fire burns some mature trees with some surviving (Lewis et al., 2022). High fire severity is where the fire burns and kills most or all trees (Lewis et al., 2022). With the severity of wildfires, there is a change in the vegetation, which can affect the food web of that ecosystem at higher trophic levels. In areas of low fire severity, lower level vegetation experiences a short term flush while in areas of moderate to high fire severity, there is a long term flush in vegetation (Lewis et al., 2022). Consequently, large herbivores that are area demanding and require a large space can choose between areas experiencing different fire severity based on their vegetation needs. To further expand, they can utilize low fire severity areas for cover and resources while using moderate to high fire severity areas for food pulses (Lewis et al., 2022). As a result of this, large carnivores can change their habitat use to mirror that of large herbivores for their own food needs (Lewis et al., 2022). Ultimately, large herbivores and carnivores may choose habitats that offer resources, and shelter and have heterogeneity in fire severity as they are useful. Going further, this indicates that allowing fire adapted forests to experience varying severity of fires can be beneficial for large herbivores and carnivores (Lewis et al., 2022). Additionally, understanding the habitat choices of animals based on wildfires can help conservation biologists when mitigating biodiversity loss due to wildfires.

Climate change can affect the stress experienced by animals such that they change behavioral responses in ways that are more maladaptive (Kay et al., 2021). For instance, areas that are increasing in temperature as a result of climate change will cause the species within them to experience physiological stress and the corresponding behavioral response would be to move to another habitat that matches their preferred temperature (Kay et al., 2021). However, moving to another habitat could result in the species becoming an invasive species that outcompete native species, resulting in the possible extinction of native species (Kay et al., 2021). The species could also decrease as a result of moving habitats if they experience increased predation or a lack of resources (Kay et al., 2021). Moreover, with climate change, there is an increase in the duration of fire seasons and there are more intense fires occurring (Kay et al., 2021). This adds to the stress that animals experience due to the sudden change in environment. The sudden change in the environment can lead to mismatches between the species occupying the habitat and the fitness those species experience in that habitat (O'Neil et al., 2020). To further explain, should the environment change but not the selection cues used by the species, the species may face a decline. Take breeding site fidelity as an example, where species will breed in the same site as it contains the resources they need and has proven beneficial for breeding in the past (O'Neil et al., 2020). This breeding site is only useful when the breeding site has stable ecological conditions, however, should the breeding site experience a wildfire, it is no longer suitable as there would no longer be any resources present there (O'Neil et al., 2020). In this case, the adaptive habit the species possess of breeding site fidelity becomes maladaptive and a mismatch occurs. This is one of the ways wildfires can affect the survival of animals following wildfires.

Animals that are specialists are more prone to the aftereffects of wildfires. According to the ecological niche theory, specialist animals perform well in ecosystems that have narrow ecological conditions as opposed to a wide range of ecological conditions (O'Neil et al., 2020). As such, they are less adaptive to other environmental conditions outside of the one they are specialized to. Therefore, when a disturbance such as a wildfire occurs, these species are greatly affected by them and often face decline (O'Neil et al., 2020).

For many ecosystems, it is unknown how biodiversity will return following a wildfire, especially since post-fire recovery is influenced by post-climate conditions that may either impair or induce vegetation recovery. One study examines the effects of fire severity and time on birds in the Mountain Ash forest regions of southeastern Australia (Lindenmayer et al., 2022). This area experiences infrequent, high severity fires that are characterized as the normal fire regime this area undergoes (Lindenmayer et al., 2022). The 2009 wildfires occurred in this region, where long-term monitoring had also occurred, allowing for pre and post-measurements following the wildfires (Lindenmayer et al., 2022). Using the data, the researchers were able to examine how the severity of wildfires impacts the occupancy of different bird sites (Lindenmayer et al., 2022). Additionally, Lindenmayer and team examined whether the life history traits of the birds are the reason for the recovery trajectories they experience (2022). It was hypothesized that there would be a strong interaction between fire severity and time period, which was demonstrated in the results. To explain, it was found that the most marked declines in birds occurred immediately following high fire severity (Lindenmayer et al., 2022). These sites would also have a rapid recovery, although the life history traits of the birds were not able to explain this phenomenon (Lindenmayer et al., 2022). This rapid rate of recovery was maintained across species and groups within the Mountain Ash forests (Lindenmayer et al., 2022). These findings provide evidence of rapid recovery for bird species from highly severe fires; however, it is possible that anthropogenic activities may impede recovery rates (Lindenmayer et al., 2022). For example, salvage logging that occurs following a fire may decrease the rate of recovery (Lindenmayer et al., 2022). Moreover, birds, which migrate, may not be the most ideal indicators of response to the fire regimes and so, recovery rates of other organisms should also be studied and examined.

Damage to Aquatic Organisms from Runoffs

Wildfires that occur close to aquatic ecosystems can alter the ecosystem. Firstly, with wildfires, there is the potential for runoff where toxins from the wildfire mix with the aquatic ecosystem (Carvalho et al., 2019). Wildfire runoff contains compounds such as metals and polycyclic aromatic hydrocarbons, which are toxic to aquatic animals (Carvalho et al., 2019). Exposure to the toxins within the runoffs was found to reduce invertebrate feeding, the biomass of fungi, and leaf litter decomposition by microbes (Carvalho et al., 2019). This disturbance to the ecosystem affects all levels within the food web of that ecosystem, from the primary producers to the apex predators. Additionally, the increase in runoff is partially due to the loss of ground-level vegetation that might typically prevent runoff from occurring by absorbing the water (Gomez Isaza et al., 2022). The lack of ground vegetation also makes the aquatic ecosystem more prone to erosion, increasing sediment runoff into the water (Gomez Isaza et al., 2022). The increase in sediments, especially phosphorus and nitrogen, results in algal growth in aquatic ecosystems (Karlson et al., 2021). The algal blooms can produce phycotoxins, physically damage aquatic organisms and deplete oxygen within aquatic ecosystems. Altogether, this impacts the survival of aquatic animals (Karlson et al., 2021).

There is also the possibility that wildfires will burn any vegetation that offers cover to aquatic systems, increasing the light that penetrates into the water. The increased UV can increase the water temperature, which consequently affects the metabolism of aquatic organisms (Gomez Isaza et al., 2022). To further explain, many aquatic organisms are ectotherms, meaning that their body temperature is reliant on and affected by the temperature of the water (Volkoff & Rønnestad, 2020). Therefore, as the water temperature increases, so does their body temperature and metabolic rate (Volkoff & Rønnestad, 2020). The increased metabolic rate will change the energy balance within aquatic and subsequently affect their behavior such as feeding and locomotion (Volkoff & Rønnestad, 2020). The relationship between feeding patterns and temperature is complex and is dependent on the behavior the animal uses to find food. For example, food detection for food intake involves chemical, mechanical and visual processes and the temperature has been shown to impact these processes (Volkoff & Rønnestad, 2020). In the rockfish species, a ten-degree increase in temperature decreases the low light sensitivity of the retina, decreasing food detection and food intake (Volkoff & Rønnestad, 2020). Furthermore, temperature affects the swimming abilities of aquatic organisms such that increases in temperature decrease locomotion speed (Volkoff & Rønnestad, 2020). Hence, the increase in light from the loss of vegetation surrounding aquatic ecosystems from a wildfire can impact the physiology of the animals within said ecosystem.

Conclusion

The aftermath of a wildfire and the effects it has on animals is quite complex. For one, there are physical injuries from the fire itself that need to be considered when it comes to caring for animals that have been in a recent wildfire. In addition, while some of these physical injuries can be easily visualized such as burns, there are physiological changes that an animal exposed to a fire may undergo, resulting in behavioral changes (Sanderfoot et al., 2021). Following the wildfire, the environment undergoes changes which result in the animals within these environments also undergoing changes. For example, some animals may face population declines due to a lack of resources or if they are specialist species (O'Neil et al., 2020). However, some species may find themselves increasing in population size, especially if the ecosystem following the wildfire becomes more suitable for them (Lewis et al., 2020). There is also the possibility of wildfires affecting aquatic organisms in the form of toxic chemicals from wildfires leaking into aquatic systems, affecting aquatic animals by changes to the food web or to their behavior (Carvalho et al., 2019; Volkoff & Rønnestad, 2020). Hence, there are several impacts on the animal population when wildfires occur and a greater understanding of these impacts can help mitigate wildfire damage or speed up recovery processes for animals following a wildfire.

References

Bosso, L., Ancillotto, L., Smeraldo, S., D'Arco, S., Migliozzi, A., Conti, P., & Russo, D. (2018). Loss of potential bat habitat following a severe wildfire: a model-based rapid assessment. *International Journal of Wildland Fire*, *27*(11), 756–. https://doi.org/10.1071/WF18072

Braun, A. C., Faßnacht, F., Valencia, D., & Sepulveda, M. (2021). Consequences of land-use change and the wildfire disaster of 2017 for the central Chilean biodiversity hotspot. *Regional Environmental Change*, *21*(2). https://doi.org/10.1007/s10113-021-01756-4

Butkus, C. E., Peyton, J. L., Heeren, A. J., & Clifford, D. L. (2021). PREVALENCE, TREATMENT, AND SURVIVAL OF BURNED WILDLIFE PRESENTING TO REHABILITATION FACILITIES FROM 2015 TO 2018. *Journal of Zoo and Wildlife Medicine*, *52*(2), 555–563. https://doi.org/10.1638/2020-0093

Carvalho, F., Pradhan, A., Abrantes, N., Campos, I., Keizer, J. J., Cássio, F., & Pascoal, C. (2019). Wildfire impacts on freshwater detrital food webs depend on runoff load, exposure time and burnt forest type. *The Science of the Total Environment*, *692*, 691–700. https://doi.org/10.1016/j.scitotenv.2019.07.265

Gomez Isaza, D. F., Cramp, R. L., & Franklin, C. E. (2022). Fire and rain: A systematic review of the impacts of wildfire and associated runoff on aquatic fauna. *Global Change Biology*, *28*(8), 2578–2595. https://doi.org/10.1111/gcb.16088

Ifaw. (2021, October 7). *How wildfires affect wildlife*. Ifaw. Retrieved December 16th 2022, from https://www.ifaw.org/journal/wildfires-impact-wildlife

Karlson, B., Andersen, P., Arneborg, L., Cembella, A., Eikrem, W., John, U., West, J. J., Klemm, K., Kobos, J., Lehtinen, S., Lundholm, N., Mazur-Marzec, H., Naustvoll, L., Poelman, M., Provoost, P., De Rijcke, M., & Suikkanen, S. (2021). Harmful algal blooms and their effects in coastal seas of Northern Europe. *Harmful Algae*, *102*, 101989–101989. https://doi.org/10.1016/j.hal.2021.101989

Kay, C. B., Delehanty, D. J., Pradhan, D. S., & Grinath, J. B. (2021). Climate change and wildfire-induced alteration of fight-or-flight behavior. *Climate Change Ecology*, *1*, 100012.

Lewis, J. S., LeSueur, L., Oakleaf, J., & Rubin, E. S. (2022). Mixed-severity wildfire shapes habitat use of large herbivores and carnivores. *Forest Ecology and Management*, *506*, 119933–. https://doi.org/10.1016/j.foreco.2021.119933

Lindenmayer, D. B., Blanchard, W., Bowd, E., Scheele, B. C., Foster, C., Lavery, T., McBurney, L., & Blair, D. (2022). Rapid bird species recovery following high-severity wildfire but in the absence of early successional specialists. *Diversity & Distributions, 28*(10), 2110–2123. https://doi.org/10.1111/ddi.13611

O'Neil, S. T., Coates, P. S., Brussee, B. E., Ricca, M. A., Espinosa, S. P., Gardner, S. C., & Delehanty, D. J. (2020). Wildfire and the ecological niche: Diminishing habitat suitability for an indicator species within semi-arid ecosystems. *Global Change Biology, 26*(11), 6296–6312. https://doi.org/10.1111/gcb.15300

Sanderfoot, O. V., Bassing, S. B., Brusa, J. L., Emmet, R. L., Gillman, S. J., Swift, K., & Gardner, B. (2021). A review of the effects of wildfire smoke on the health and behavior of wildlife. *Environmental Research Letters, 16*(12), 123003–. https://doi.org/10.1088/1748-9326/ac30f6

Venn-Watson, S., Smith, C. R., Jensen, E. D., & Rowles, T. (2013). Assessing the potential health impacts of the 2003 and 2007 firestorms on bottlenose dolphins (Tursiops trucatus) in San Diego Bay. *Inhalation Toxicology, 25*(9), 481–491. https://doi.org/10.3109/08958378.2013.804611

Volkoff, H., & Rønnestad, I. (2020). Effects of temperature on feeding and digestive processes in fish. *Temperature (Austin, Tex.), 7*(4), 307–320. https://doi.org/10.1080/23328940.2020.1765950

An Unwelcome Visitor

Conclusion

Fire is a complex entity. Born as a building block of life, it can quickly become the catalyst for the end. Wildfires wreck havoc on the delicate cycles of life, affecting human and animal life unlike any other natural disaster. The increase in atmospheric carbon dioxide has created a feedback loop of sorts, resulting in an increased number of fires across the globe, the aftermath of these fires affecting all species on earth. We hope that you, the reader, leaves this book with an increased appreciation and knowledge of the impacts that wildfires have across the world. From the boreal forests of North America to the dry bush of Australia, the effects of wildfires will likely touch everyone as wildfires have been predicted to get worse in future years with limited options to stop the increase. The world is becoming increasingly prone to wildfires, and the astounding amount of land scorched and lives affected is only projected to rise. Smog, smoke, and flame itself can all have disastrous impacts on life, and the global increase in carbon emissions only serves to exacerbate an already delicate system. As with any natural disaster, the effects are not felt equally, and the massive toll that wildfires exhaust affects each country and each individual differently. Through a robust understanding of the carbon cycle, global investment, as well as developed restoration strategies and prevention efforts, perhaps we can stem the impact of these disasters and take action to protect the Earth's climate and preserve the natural world for future generations. This book seeked to create a timely, relevant, and accurant compilation of the scientific evidence of wildfires and their growing rates in an easy-to-consume manner to help drive discussion and the desire for change. While it is hard to foresee what the future holds, as fires can be started in an instance, one thing remains clear: we all have a role to play in preventing wildfires.